MW00632579

# Urban Farmers

*The Now (and How) of Growing Food in the City*

gestalten

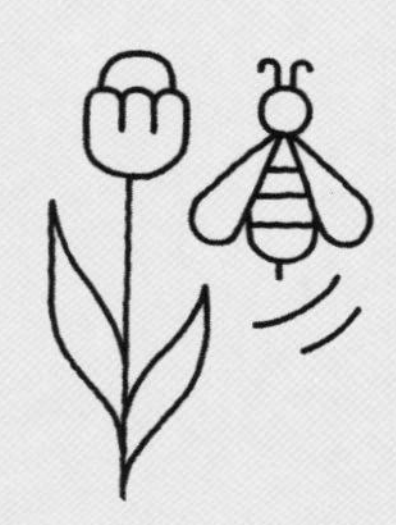

# CONTENTS

# The Joys of Growing Food in the City

It's a sun-dappled summer morning and the air smells of rosemary and dirt. A toddler stumbles after chickens and ducks in vain, while her parents pick strawberries and fresh herbs as they chat with a farm volunteer. Behind them, skyscrapers cast their shadows over this little green haven in the middle of the city...

It might be unthinkable to some, but a scene like this is increasingly within reach for city dwellers around the world hungry to interact with nature and with each other. Among the many advantages of urban farming, a truly remarkable one is its ability to connect human beings to the natural world around them. The concrete jungles of modern cities might seem hostile to nature, but they are full of natural life. And more often than not, it is in our cities that children first get acquainted with the outside world.

According to the United Nations, 55 percent of the world's population lives in urban areas, a figure that is projected to increase to 68 percent by 2050. While it's true that urban development has wiped out most signs of agrarian legacy in cities across the world, today there are more options than ever before to grow food in urban settings. The reasons behind this surge in urban farming are multifarious: from addressing food insecurity to focusing on biodiversity, urban farmers have a unique opportunity to touch on many disciplines at once and become dynamic forces of positive change in their communities.

Why shouldn't our cities become wildlife corridors and refuges for biodiversity? Green spaces, from parks to green roofs, are hugely advantageous in myriad ways—they reduce air pollution and offset global warming by lowering inner-city temperatures and preventing the urban heat island effect. Green spaces also help the conservation of species in the city by supporting biodiversity and wildlife habitats, including those for pollinators, a key link in the food chain.

It is estimated that urban agriculture can provide 15 to 20 percent of the world's food supply. Urban farming has the potential to improve food access locally, which is especially relevant in underserved communities where nutritious food is scarce. It also offers the opportunity for a culturally diverse diet. Since 1970, the number of people living in a country other than where they were born has tripled, and today 3.5 percent of the world's population are international migrants. Culturally diverse foods represent an opportunity for different communities to maintain strong ties to their homelands, and to reconnect to the food ways of their ancestors.

And in a world hit by a pandemic, urban farming spaces are fast becoming oases of fresh air, gathering places for urbanites to engage in physical activity and interact with each other from childhood to senectitude, as part of a healthy outdoors lifestyle. Urban agriculture, framed in the wider context of multipurpose urban green spaces, can have a positive impact on an individual's mental well-being and also improve quality of life. Proximity to urban greenery is crucial to enhance the scope of good mental health and reduce stress-related forms of mental disorders, which in Europe alone are one of the biggest public health challenges.

The pandemic has encouraged many to focus more carefully on their well-being, whether through a healthier diet (dozens of studies show that those involved in growing food are more likely to eat fresh foods such as vegetables and fruit), more exercise, or more quality time with loved ones—as long as it's allowed—without leaving their neighborhoods.

There are as many reasons to start growing food as there are urban farmers. Working with photographer and co-editor Valery Rizzo, our goal with this book is to showcase the vast universe of urban agriculture in different parts of the world, revealing what makes each of the farms and projects featured unique. Some run their projects as innovative, for-profit start-ups where the latest technologies rule, while others pursue a nonprofit approach based on alleviating systemic structural issues affecting their communities, from food insecurity to the inescapable lack of opportunities that people living in poverty regularly suffer. Many convert their farms into temporary educational havens, providing a journey of discovery for city dwellers curious about where the food on their plates comes from.

One of the shared challenges of growing food in the city, no matter where, is the limited space available, which often informs what farming methods can be used. But technology has considerably expanded the choices available. Today we can cultivate food on rooftops or produce microgreens in vertical, hydroponic farms underground. In these pages, we have included a comprehensive overview of different farming techniques, with an aim to inspire and provide a base of knowledge to better understand the idiosyncrasies of growing food in urban settings.

Throughout these chapters, which range from case studies to profiles and guest essays, our aim is to showcase the diversity and wealth of urban farming projects in different parts of the world and the many possibilities available for anybody to join the movement, regardless of their level of experience or the scope of their ambition. If we manage to pique your interest, there is also practical information designed to help you take the first small steps into the immense cosmos of urban farming.

Most of the savoir faire on which sustainability-minded farmers build their urban farming projects has been passed down through generations. In their hands, that agrarian knowledge keeps evolving, as it has been for millennia. And many would agree that the best way to learn is to roll up your sleeves and join in.

*Mónica R. Goya is a journalist and photographer covering farming, travel, food, and the environment. Her personal projects explore the culture of working the land and the intersection of human rights, food politics, and sustainability.*

# History and Pioneers

*Farming in cities is hardly new: its history stretches back millennia*

According to the UN, nearly 70 percent of the world's global population is projected to live in urban settings by 2050. While growing food in cities isn't new, the layers of significance associated with it have changed over time and across different countries. During the era of industrialization, urban gardening was a subsistence tool for the urban poor. That's still the case in many locations today, but it has also become popular among those who don't grow food exclusively for sustenance, but practice it as part of a healthy, environmentally friendly lifestyle. Growing fruit and vegetables in cities has also become a way for food justice activists across the world to fight for a more equal food system.

Recent trends are only the latest in a long history of significant examples of urban farming dating back to ancient times. Throughout the arid regions of what is now modern-day Iran, ancient Persians invented the *qanat* system, an ingenious network of subterranean aqueducts that provided fresh water for purposes such as irrigation. Water was conducted from the hills to the towns through this underground system, protecting it from the scorching sun. Eleven *qanats*—the oldest dating back around 2,500 years—are now included on the UNESCO World Heritage List.

*The layers of significance associated with growing food in cities have changed over time and across different countries.*

With over 21 million citizens, Mexico City is one of the largest metropolises on Earth. It was built on an ancient lake bed, where pre-Hispanic civilizations developed highly sophisticated systems of food production,

such as the man-made floating island vegetable gardens known as *chinampas*. These ancient farmlands have to some extent survived in the district of Xochimilco, and are a living reminder of the Aztec city of Tenochtitlan; they were designated a UNESCO World Heritage Site in 1987. The area houses 39 square kilometers (15 square miles) of protected wetlands where traditional farming is still alive. Lucio Usobiaga (see pages 160–165) is leading efforts to conserve the *chinampas* with his project Arca Tierra.

In France, the *Hortillonnages,* small floating gardens built on reclaimed marshland, are a symbol of the town of Amiens. Covering 300 hectares (740 acres), these market gardens are interwoven with 65 kilometers (40 miles) of small waterways and have been cultivated since the Middle Ages. In nineteenth-century Paris, when horses were the main means of transportation, horse manure was transformed into precious compost that would support the production of thousands of tons of salad crops year-round.

In the agitated first decades of the twentieth century, necessity made the victory gardens movement popular across the world, from North America to Australia. In Britain during the Second World War, the British Ministry of Agriculture launched the "Dig for Victory" campaign in 1939 to counter declining food imports from abroad. It was so successful that allotment numbers grew to 1.7 million in only three years. Even London Royal Parks, including Hyde Park, set land aside to grow vegetables as part of the campaign.

*In the agitated first decades of the twentieth century, necessity made the victory gardens movement popular across the world.*

Decades before the victory gardens sprouted up, the "Schreber movement" emerged in Germany. It was named after Moritz Schreber, a Leipzig physician who championed the idea of reserving green

*The growing of food in cities is an ancient tradition. In Amiens, France, the* Hortillonnages *(floating gardens) have been cultivated since the Middle Ages (p. 7). During the Second World War, children worked in school victory gardens in New York City, while women joined the "Dig for Victory" campaign in the U.K. (top and p. 9 top). In many German cities, little plots of land known as* Schrebergärten *offer city dwellers relaxation and the chance to tend to nature (p. 9 bottom left and right).*

spaces for children to play and spend time outdoors during the era of industrialization. Over time, these green areas scattered across the country evolved into allotments, or *Schrebergärten*. Usually located on the outskirts of cities, today they are particularly visible from trains, instantly recognizable for their fenced-off plots and colorful little huts. Considered a pastime for conservative-leaning retirees in previous decades, these days interest in them is growing among the young—in cities like Berlin, there is a three-to-five-year waiting list to rent one. *Schrebergärten* are managed by tenants' associations, and while the socioeconomic backgrounds of gardeners might have changed, the rules continue to be strict.

*The Green Guerrillas were pioneers in harnessing the power of community gardening to transform neighborhoods.*

The Green Guerrillas movement, a pioneering example of harnessing the power of community gardening to transform entire neighborhoods, didn't exactly play by the rules at first. Founded in New York City in the early 1970s by artist and activist Liz Christy, the movement started by throwing seeds, or "seed bombs," over the fences of vacant lots. The group also planted sunflower seeds in some of the city's busiest street meridians and placed flower boxes on the windowsills of derelict buildings. Increasingly ambitious plans followed, culminating in the conversion of a vacant lot into New York City's first community garden. The Bowery-Houston Community Garden (now the Liz Christy Bowery-Houston Community Garden) sparked a movement that continues to grow; today, there are hundreds of community gardens in the city.

*Empty supermarket shelves revealed the fragility of current food systems in the early days of the Covid-19 pandemic.*

In the early days of the Covid-19 pandemic, empty supermarket shelves—once an unimaginable sight in the developed world—revealed the fragility of current food systems. With the global food supply chain thrown into disarray, stockpiling became widespread and grocery stores lacked essential products such as flour and eggs. To an extent, the latter might have fueled the revived push for locally produced foods.

A considerable number of urban farms offering CSA programs, which provide households with a box of locally grown fruit and vegetables, have reported increases in demand that tripled or quadrupled their pre-Covid-19 numbers. Vegetable seed sales soared during the first months of lockdown, to the point that seed companies struggled to meet demand across the Western world. There's little doubt there is novel interest and legitimacy in growing food among urban residents of all socioeconomic backgrounds, and that increasing numbers of urbanites are digging in their back gardens or filling their balconies with plant pots, sowing seeds and re-thinking their living spaces to include more greenery. The question is, will the attention last?

*There are many different reasons to grow food in the city. In the late 1940s, when the Soviets constricted supply routes into West Berlin during the Berlin Blockade, inhabitants grew their own vegetables and tobacco in allotments on Berliner Strasse (p. 11 top). In New York City, Green Guerrillas volunteers prepare mulch in the spring for the Liz Christy Community Garden (p. 11 bottom left), while Brooklyn Grange Rooftop Farm (see p. 226) founders Gwen Schantz, Anastasia Cole Plakias, and Ben Flanner run a successful rooftop farming company (p. 11 bottom right).*

99
5509

# Reviving Neglected Neighborhoods with Bountiful Crops and Green Space

*Urban farm MUFI, located in Detroit's North End, connects residents with agriculture and each other*

Founded in 2011, nonprofit organization MUFI (Michigan Urban Farming Initiative) focuses on community engagement, education, sustainability, and food security. In growing fresh produce—fully powered by volunteers—and offering it to underserved communities free of charge, they provide a healthy, local, and affordable alternative to the ubiquity of industrial food.

MUFI was co-founded by Tyson Gersh, whose interest in agriculture grew while he was a psychology student at the University of Michigan-Dearborn. He first became familiar with Detroit's North End community in his early twenties while working as a research assistant for the Urban Community Oral Health Project. During that time, Gersh saw first-hand the devastating consequences of food insecurity and how inequalities within the food system are linked to structural disparities and socioeconomic status. Together with his university friend Darin McLeskey (who is no longer involved), he founded MUFI with the aim of addressing the issues he had witnessed.

Detroit is a complex place. Once the prosperous cradle of the automobile industry in the U. S.—its nickname is still Motor City—in 2013 it became the largest municipality in the history of the United States to file for bankruptcy, and has lost 20 percent of its population in the last decade alone. Neglected neighborhoods are not a rare sight: almost 20,000 abandoned houses have been knocked down since 2014, and inequality is part of everyday life, with more than a third of the population living below the poverty line. Initiatives to improve the neighborhoods walk a tight line between enhancing and gentrifying—an issue that is complicated, and that urban farming projects (including but not limited to MUFI) must navigate consciously and honestly in order to truly succeed in benefiting the community.

One of MUFI's top priorities is to address that rampant inequality. "Our mission is technically to engage people in urban agriculture, but effectively our work accomplishes so much more," says Gersh, the farm's executive director. "I think the value that our organization brings to the community has a lot of different parts to it: food security, of course, and also having a big volunteer force in the neighborhood to help existing residents with property maintenance and community beautification. We also provide an outdoor space to socialize in, where residents can be effectively grocery

shopping and interacting with people at the same time, which has become increasingly important, especially during Covid-19."

On the MUFI farm around 9,000 kilograms (20,000 pounds) of produce is grown annually, and in spite of a considerable decrease in volunteers during the pandemic (the average number of volunteers in a typical year dropped from 2,000 to 200), they have remarkably continued to tend the same amount of fresh produce. "It's been a lot of hard work," says Gersh.

The fields on the farm are divided by zones with an outstanding diversity of crops. In the 0.5 hectare (one acre) of land they have in production, over 300 varieties of vegetables are grown, including more than 20 varieties of tomatoes and peppers and a dozen different collard greens, which are among the most popular crops. After listening to residents, they incorporated culturally diverse foods important to their community.

The space also includes a high-density orchard of 200 fruit trees, producing apples, pears, cherries, plums, and peaches, and a sensory garden that encourages interaction with agriculture. "The sensory garden is a kids' space where the plants' inventory has been curated based on different plant qualities," says Gersh. "So there's a section of different-textured plants—some are spiky, some are soft, some are rubbery—and another section focused on visual stimulation. The idea is to create a space that engages the five senses deliberately, and the goal is for the public to interact with it, not necessarily by eating the produce."

MUFI's farming method is intuitive: the community sets the pace. "We are organic, effectively, but we haven't gone through the certification process," says Gersh. "We chose organic because that's what people said they wanted." Their distribution model has shifted recently: whereas previously they offered volunteer-powered bike deliveries, now, due to a scarcity of volunteers, they have adopted an on-demand approach in which local residents can go to the farm and have direct access to anything that is being grown there at no cost. "Anybody that wants fresh produce can come to the farm and we give it to them—it's all free," says Gersh. "It's really important that it's free for our community; it's something that people who live in this neighborhood have consistently expressed an interest in." They are working toward a self-serving model in which each person can harvest

*1] MUFI is a nonprofit initiative that seeks to engage members of the Michigan community in sustainable agriculture (p. 13)*
*2] Work on the farm is powered by local and corporate volunteers (p. 14 top)*
*3] An assorted variety of crops are grown, including herbs (p. 14 bottom) and tomatoes (top)*

ONE WAY
SATURDAYS
CLOSED
TODAY
OPEN

Michigan Urban Farming Initiative

mobile mini

anything they fancy themselves, but since plants can be easily damaged unintentionally, they haven't yet found an efficient way to make it work.

Future projects include a healthy café, which Gersh hopes will make the venture financially viable. "It's really a fantastic public space, but we haven't capitalized on that in any way. I think the café will bring more people to the farm, give them another reason to be here," he says. "The aim is to make it a self-sustaining model that doesn't require charity."

Prioritizing green space to enhance quality of life is one of the ideas behind the concept of "agrihoods," defined as residential communities built purposely around a working urban farm. MUFI is often credited as the first sustainable urban agrihood in the United States. In addition to the café, they are hoping to develop a Community Resource Center (CRC) at a property located across the street that MUFI purchased at a tax auction. The CRC will incorporate an event and meeting space and will offer a range of educational programs, as well as a nonprofit incubator space. In the meantime, MUFI will continue to provide fresh produce to local residents, as well as making their green space available for the community to enjoy.

*4] MUFI is located in Detroit's North End neighborhood (p. 16)*
*5] MUFI stands for Michigan Urban Farming Initiative (p. 18 top)*
*6] Farm tools help volunteers with the workload on the farm (p. 18 bottom left)*
*7] Colorful fields are also home to a variety of flowers (p. 18 bottom right)*
*8] MUFI is transforming an adjacent property into a community resource center (top)*

# Urban Farming from Scratch to the Brooklyn Rooftops

*On being a pioneer and building a new commercial farming project from scratch*

BY URBAN FARMING ENTREPRENEUR ANNIE NOVAK

I fell into farming because I like to eat chocolate. An autodidact by nature, I followed a West African cacao farmer deep into his orchard, curious to see how chocolate could grow on a tree. That was almost 20 years ago, and I never looked back. In the decades since, my inquisitiveness has not flagged, nor have the rewards plants give.

In 2005, freshly graduated and ripe with green ambition, I jumped into New York City's urban food and farm scene, convinced I could help steer the hearts and minds of all eaters toward the local, the sustainable, and the just. I took an internship at the New York Botanical Garden. I apprenticed under Tri-state area farmers. To hone my skills, for several winters in a row I took all the money I saved by growing my own food and flew to the southern hemisphere, finding food and farming jobs by socializing at local markets. Relying on the kindness of strangers, I learned to fish, milk, and butcher animals.

Inspired by similar programs out west, I founded a nonprofit, Growing Chefs: Food Education from Field to Fork. At 23, I put up a website, gave myself the title of "director," and set out to create the job I wanted. With a part-time teaching staff of gardeners, nutritionists, educators, and cooks, I built school gardens and taught cooking and gardening classes informed by math and science. While mastering vegetables, I discovered soil. From soil grew a love for fungi. Fungi made me fall for trees. I cached data like a magpie and counted myself wealthy.

In the fall of 2008, another young self-starter, Ben Flanner, approached me at one of my farmers' market gigs and asked if I would look over his plans to start a green roof farm atop a warehouse on Eagle Street in north Brooklyn. The term was alien to me. I felt skeptical about the growing media, confused how dirt, displaced, would host the right microbes and decomposers. Soil, after all, is a farmer's principal investment: seasons of building out a good biological community yield healthier, more nutritious crops. But the following April, with the support of the building's owner, Broadway Stages, Goode Green installed the green roof and I set to planting peas.

Ben went on to found Brooklyn Grange while I stayed on at Eagle Street, expanding into chickens, bees, and—briefly and unbeknownst to me illegally—goats. I started an apprenticeship program of my own, which over 10 seasons (and counting) has hosted over 125 growing green thumbs. As they move on to their own projects, this

new generation of growers is benefiting from an increasing number of initiatives and legislation at the city, state, and federal levels in support of urban farming, green roofs, green spaces, and food policy. However, I suspect that the curiosity and enthusiasm that brings them up the stairs to join me on the rooftop farm is perennial.

A chocolate farmer once astonished me with the jittery deliciousness of a single cacao bean. Now I stand three stories up in the air and watch the faces of my apprentices and our visitors open with joy and surprise when they taste the fulsome sweetness of a fresh carrot or the sharp bite of their own radish, planted the month prior. I've learned so much about growing things since setting off blindly all those years ago. But chief among these is the fact that productive and rewarding urban agriculture is within the reach of anyone determined to make things grow.

*Annie Novak is the head farmer and co-founder of Eagle Street in Brooklyn, the first commercial green roof farm in the United States. She is also an author, educator, and speaker on the benefits of urban agriculture.*

# Edible Urban Landscape Design with a Sustainable Approach

*Founded in Paris with multiple projects sprouting up across France, Topager creates edible urban landscapes*

Topager was founded in 2013 by Frédéric Madre, an expert on biodiversity integrated into the urban landscape, and Nicolas Bel, a gardener and urban farming expert. Its aim is to provide a sustainable and innovative approach to edible and urban landscape design.

Madre, who is involved in dozens of farming projects across the country, is adamant about their multiple benefits. "Urban open-air agriculture and agroecology make it possible to create local jobs, fight against urban heat islands, make energy savings in buildings, improve rainwater management, store carbon dioxide, and increase the well-being of urbanites," he says.

One of the company's most well-known projects is the Opéra 4 Saisons rooftop farm, which sits above the Opéra Bastille building in the heart of Paris's 12th arrondissement. The edifice, which was inaugurated in 1989, has a long history of controversies and rejection by the general public—but this farm presented an opportunity to use part of it for a new and exciting purpose. The project was installed in 2018, two years after it won a call for proposals by the Parisculteurs campaign, an initiative that aspires to add 100 green hectares (247 acres) to the city's existing rooftops, walls, and facades, with approximately one-third devoted to urban agriculture.

Today, Opéra 4 Saisons is considered one of the largest urban farms in Paris, covering 2,500 square meters (27,000 square feet) over four rooftops, with a further 2,000 square meters (22,000 square feet) spanning the facades. Located on Place de la Bastille, where the infamous prison once stood, urban farmers can take in expansive views over the city and up to Montmartre, with the gilded Spirit of Freedom atop the July Column in the foreground. Sometimes, they can even work to the enrapturing sound of opera rehearsals happening downstairs.

A significant portion of the production at Opéra 4 Saisons focuses on light plants such as aromatics, or uncommon edible flower varieties including pineapple sage and oyster leaf. In addition, they also grow heirloom tomatoes and salad leaves. In total, they harvest around two tons of vegetables per year. "We sell veg boxes every week to employees and musicians of the Opéra," Madre says. "It's a one-year membership scheme and the boxes have a value of €15 ($18) per week. We also grow produce for a gourmet restaurant in the neighborhood."

Topager designed the farm in a way that considers biodiversity as a whole. Facilities that support wildlife, such as nesting boxes or small ponds, have been incorporated into the Opéra's rooftop design with the use of mostly natural and local materials, including wooden vegetable plots or plants from local nurseries.

*"Urban open-air agriculture and agroecology make it possible to create local jobs, fight against urban heat islands, improve rainwater management, store carbon dioxide, and increase the well-being of urbanites."*

Additionally, hops—including heritage French varieties such as Tardif de Bourgogne—are grown across the facades, providing valuable bioclimatic advantages. "Hops are plants that grow quickly in spring and cover the facades with their large leaves, so they refresh the building in summer, and in winter the plants are no longer present—we cut the lianas in autumn so the building can benefit from solar gains," Madre says. An artisanal micro-brewery located in the premises, under the Opéra's roof, is also part of Topager, who are developing a brand called TopHop and plan to open a larger brewery in northern Paris.

Topager's lease of the Opéra's rooftop lasts until 2030, with the possibility of an extension to 2038, a maximum total of two decades. The initial investment of €350,000 ($426,000) was undertaken by both Opéra itself and Topager. "The Opéra had to redo its roofs anyway, so the investment was planned," explains Madre. "If the project does not work over time, they will have beautiful green roofs and we will sow local flowers."

Madre gives credit to his passionate 25-employee team, five or six of whom work at Opéra 4 Saisons alongside around 10 volunteers brought in by The Gardeners of the Opéra association. "Our employees are empowered and make their own decisions at their level," says Madre. "The crop manager of the Opéra 4 Saisons project, Marie Carcenac, chooses the varieties that are replanted each year and establishes her cultivation plan. We just give her suggestions for certain species and she also listens to members of the veg box scheme and to the restaurant's cooks."

*1] Harvest day atop the Opéra Bastille (p. 23)*
*2] Axelle harvests romaine lettuce, included in the weekly baskets available for Opéra employees and local residents (p. 24 top)*
*3] Crop manager Marie Carcenac explains to Joséphine how to tie the bundles of basil after harvesting (p. 24 bottom)*
*4] High added-value crops like these edible cosmos flowers are also grown at Topager (top)*

While Madre and fellow co-founder Nicolas Bel are in charge of strategic decisions, essential for them is the taste and quality of what is produced and the cultivation method, based on agroecology—meaning they don't use chemicals or any non-organic treatments; they produce compost on-site to fertilize and pay attention to associations of plants that balance each other.

> *"Hops grow quickly in spring and cover the facades with their large leaves, so they refresh the building in summer, and we cut the lianas in autumn so the building can benefit from solar gains."*

Previously, the stunning rooftop city views could only be enjoyed by visitors on the rare occasions when events took place. That was about to change before the Covid-19 pandemic broke out. "This project is not profitable so far, but we believe we can achieve an economic balance by hosting visits and workshops. Because of Covid-19, we were not able to launch this initiative when we planned, but we still launched an association for people who want to get involved as volunteers, and we also welcome journalists and small groups of people interested in the project."

In their quest for more inclusive, livable cities in the context of climate change, Topager believes that developing farming projects more accessible to all in France—such as food plots on the rooftops of social housing complexes—is one of their most important challenges. Madre is optimistic about the future of urban farming. "I think a bright future lies ahead, especially after the Covid crisis and considering the growing environmental concerns about climate, biodiversity, and urban sprawl," he says. "We particularly believe in urban agroecology that is socially inclusive, rather low-tech but with a lot of scientific knowledge, as a way of providing efficient ecosystem services such as rainwater management, energy savings, and fighting against heat islands."

*5] Topager project Opéra 4 Saisons is located on the rooftops and facades of the Opéra Bastille (p. 26)*
*6] Washing the harvested greens for the weekly baskets. Produce is harvested the same morning it is distributed (p. 28 top)*
*7] Harvesting edible nasturtium flowers, popular with local chefs (p. 28 bottom left)*
*8] Lucas busy composting. Hops including the rare Tardif de Bourgogne, scale the facade behind him (p. 28 bottom right)*
*9] Axelle harvests green beans on the Opéra Bastille's rooftop (top)*

# A Lush Space for Biodiversity and Community

*Set in a formerly vacant lot in Far Rockaway, Queens, Edgemere Farm is run by passionate volunteers*

Edgemere Farm was founded in 2013 by David Selig and Matthew Sheehan as part of New York City's Gardens for Healthy Communities program, framed within wider public policy actions to tackle obesity. What was then an abandoned city-owned lot by the beach in the Edgemere neighborhood of Far Rockaway, Queens, has been converted into a lush 2,000-square-meter (22,000-square-foot) space where well-kept rows of sustainably grown vegetables, flowers, and herbs thrive.

"The farm has a mission of education, food justice, and increasing capacity for both the production of food and the processing of food waste on the Rockaway Peninsula," says Mike Repasch-Nieves, who, together with his wife, Vanessa Seis, took over the farm's management at the end of 2019, after Sheehan's retirement. "Providing the surrounding Edgemere community access to fresh produce and education is a core part of what we do."

Run by volunteers under a not-for-profit model, there are between 10 to 20 regulars, plus others who come and go, rotating throughout the season. "Our gate is open to everyone, and anyone is welcome to come and get their hands dirty with us for as long or as little as they like," says Repasch-Nieves. "Every person who volunteers with us is invited to take home produce, so it's also a way for anyone who might have difficulty financially to be empowered."

*"Far Rockaway residents don't have consistent access to fresh fruits and vegetables or healthy food options. The farm was founded as a direct response to that."*

The farm developed in the wake of the destruction triggered by Hurricane Sandy. "Among other things, this really brought to light the fact that Far Rockaway is a food desert—now federally labeled—meaning that residents don't have consistent access to fresh fruits and vegetables or healthy food options. The farm was founded as a direct response to that," says Repasch-Nieves. He feels it's important for the farm team to demonstrate how productive a relatively small plot of urban land can be; the amount of food waste they process into useful energy—in the form of compost, fertilizer, or chicken feed—would otherwise likely end up in landfill.

Camping Aeroport
Quebec

When co-founder David Selig, who is still involved in the farm, was awarded the NYC Parks Department contract to operate the boardwalk food concessions in Rockaway, he committed to operate the food businesses with high standards in waste management, including the composting of all organic matter. Today, the composting program is a vital part of Edgemere Farm. They accept household food scraps for composting any-time, and they partner with local restaurant Cuisine by Claudette to collect all of its food waste daily, along with spent grains collected weekly from a local brewery that are fed to Edgemere's chickens.

> *"The farm has a mission of education, food justice, and increasing capacity for both the production of food and the processing of food waste on the Rockaway Peninsula."*

"The chicken program has 50 to 60 chickens processing 54 tons of organic matter every 12 months," explains Selig, who focuses mainly on the chickens, bees, and composting. "The outputs of this waste processing include almost 1,000-dozen extra-rich eggs, 15 cubic meters (20 cubic yards) of nutrient-rich soil/compost, and 3 cubic meters (4 cubic yards) of potent chicken excrement. All the outputs stay in our immediate community, promoting health through incredibly nutritious eggs, as well as fertilizing the growth of our neighbors' healthy food and flora with fertile soil." Bees also help with production on the farm, from pollination to producing honey that is sold in the summer months at the farm stand.

Climate and seasonality determine what can be grown in worthwhile quantities within the limited space. The core crops include lettuce, kale, carrots, tomatoes, zucchini, figs, and berries. "We also have a big Jamaican, West Indian, and Caribbean population in our neighborhood, so we grow a lot of things like Jamaican *callaloo*, scotch bonnet peppers, bitter melon, okra, collard greens, taro, and Malabar spinach—also known as *callaloo* in Guyana—that you don't often see grown around here, but are staples in the various cuisines of many in our community," says Repasch-Nieves. Furthermore, Babajide Alao, owner of TheCradleNYC (see p. 206), a local restaurant offering West African cuisine, is

*1] Farm managers Mike Repasch-Nieves and Vanessa Seis at Edgemere Farm, Far Rockaway, Queens (p. 31)*

*2] Babajide Alao harvests raspberries for TheCradleNYC, his local restaurant featuring West African cuisine (p. 32 top)*

*3] Repasch-Nieves and Seis harvest rainbow chard and eggplants for the Friday farm stand (p. 32 bottom)*

*4] Co-founder David Selig holds an Ancona chicken, a breed originating from the Marche region of Italy (top)*

*5] Repasch-Nieves washes and preps harvested scallions (p. 34 top)*
*6] Harvested purple top and hakurei turnips (p. 34 bottom left)*
*7] Harvested zucchini flowers, squash, and jalapeño peppers (p. 34 bottom right)*
*8] Seis selects and harvests jalapeño peppers (top)*

swiss chard

now growing food including Nigerian crops in a section of the farm.

Repasch-Nieves was a latchkey kid who ate a lot of junk food growing up. He says he started to think about where food comes from in college. After a decade touring the world playing guitar in an indie band, he moved to Rockaway from Brooklyn in 2018. "Shortly after moving, some friends invited us to come volunteer at the farm, and that was it," he says. "What started as casually volunteering quickly turned into a much bigger commitment." When co-founder Matthew Sheehan asked Repasch-Nieves and Seis to get more involved, they didn't know he was planning to retire. Sheehan and Selig mentored them, and the young couple have been running the farm since late 2019.

*"Having the gate open and having people from the neighborhood wandering in, people getting involved, there is something really special and important about that."*

They see resilience as an important skill for outdoor farmers. "Due to weather and heat, last season was a really rough one in terms of diseases and pests," says Repasch-Nieves. "Since we're growing everything here without the aid of chemicals or pesticides, we have to be crafty. There are no shortcuts. You learn pretty quickly that you just can't get attached to anything. You put all this time and energy into every single seedling, every plant, and then something happens and it all goes out the window. It's heartbreaking, but you have to just shrug, roll your sleeves back up, and keep going since there's only a finite amount of growing time each season and you need to make the most of it."

Produce is available to buy from the farm stand on Fridays and Saturdays (aside from the colder months when the farm closes), and sometimes customers can do the groceries to the sound of music: Edgemere hosts live music sessions, from bachata to string quartets.

"The value that an outdoor space like this provides for the community is not tangible," says Repasch-Nieves. "Having the gate open and having people from the neighborhood wandering in, people getting involved, there is something really special and important about that. It's hard to quantify—you don't have metrics to define it—but it's one of the biggest services that a space like this can really provide."

*9] Micca bundles and preps an assortment of fresh herbs for the farm stand (p. 36 top)*

*10] A selection of the fresh produce available (p. 36 bottom left)*

*11] Honey harvested from a single hive at Edgemere Farm. The jars on the left contain honey harvested in early August; those on the right contain a late-June harvest (p. 36 bottom right)*

*12] Louise ties bouquets for sale at the farm stand (top)*

# Community and Cultural Preservation

*Seeds form part of a community's soul, nourishing our bonds to ancestral cultures*

By improving access to green space and to healthy, culturally diverse foods, urban farms and community gardens offer an opportunity to foster community-building activities in the heart of urban centers. Having a place in the city where food is grown offers a chance for neighbors to connect with each other and with the earth. Additionally, the often intergenerational nature of these spaces provides a unique occasion for cross-cultural learning and cultural preservation, including the conservation of rare seeds important to different communities.

Prior to industrial agriculture, farmers would save the best seeds every season, an intuitive choice based mostly on observation and personal taste. Season after season, century after century, seeds would be preserved according to factors including flavor and resistance to pests. In this way, seeds became increasingly resilient by adapting to local conditions through time. Preserving seeds is an ancestral practice that dates back for millennia and, until the last century when seed companies and industrial agriculture emerged, it was an indispensable step to planting new crops the following season. That interaction between nature and food-producing communities resulted in a flourishing agricultural biodiversity that provided resilience and a precious collective inheritance, passed down through generations.

According to the Food and Agriculture Organization of the UN, vegetable crops worldwide lost 75 percent of their genetic diversity between 1900 and 2000. Truelove Seeds is a small seed company that grows, preserves, and sells open-pollinated, rare seeds. They work together with food justice and community-minded farmers toward the same goal: preserving culturally significant seeds. "People think a lot about animals at risk of going extinct, but not a lot about

varietals that are in danger of going extinct," says Owen Taylor, who founded the company in 2017. "Similar to the dinosaurs, once we lose those varieties, there's no getting them back. It means a loss of lineage, of connection to our ancestors. The loss of those varietals means there is that much less genetic diversity that can help us as our climate changes. We have more floods, more droughts, more heat, and it's important to have as much genetic diversity as possible, so that we can find that one variety that will thrive in the changing conditions."

Before founding Truelove Seeds, Taylor worked for four years at the Roughwood Seed Collection with food historian William Woys Weaver, who became his mentor. The private seed collection, which was started in 1932 by Weaver's grandfather, H. Ralph Weaver, now comprises nearly 5,000 varieties of heirloom food plants. "I thought of all these seeds disconnected from their people in his seed library," says Taylor. "In my 15 years in the food justice movement, as part of our food sovereignty work we talked about people growing affordable and accessible but also culturally important seeds. I thought I could help to focus on the culturally important seeds through seed keeping."

> *"People think about animals at risk of going extinct, but not a lot about varietals that are in danger of going extinct."*
> *—Owen Taylor, founder of Truelove Seeds*

Of Irish and Italian ancestry, Taylor grows varieties that are linked to his heritage and asks his collaborators to choose seeds that are meaningful for them. "I grow my southern Italian and Irish ancestral varieties, and the people that work here keep their ancestral seeds too," says Taylor. "I find it really valuable personally, as a White person in America, to delve into my culture and my history in a way that we were not taught to as children. My Italian and Irish ancestors in this country were given many rewards for forgetting where they came from, for forgetting their language, their peasant ways. Trading their cultural experience for the American experience brought them many economic and cultural rewards, but that meant that we lost their

*Saving seeds is an ancestral tradition and central to the work of farming communities. At Truelove Seeds in Philadelphia, seeds are collected from dried okra pods (p. 41 top and bottom left), and for a plentiful seed harvest (p. 41 bottom right),* Gbure *(waterleaf) is harvested for seed and aji amarillo pepper seeds are set out to dry (p. 39). St. John Cantius Parish Community Garden is one of over 60 community gardens supported by East New York Farms! (top). UCC Youth Farm interns Ayman, Taveya, and Tyjahnea harvest collard greens (p. 38).*

languages, food ways, rituals, and traditions, in exchange for entering the privileges of whiteness. So, for me, looking back to where my people came from and what sustained them is a way of healing that loss and also challenging the myth of whiteness that has been created in the United States and beyond."

The Irish Lumper potato is a crop that illustrates Taylor's approach: through that variety, he creates a narrative that touches on colonialism, Ireland's Great Famine, the meaning of poor countries growing food for rich countries—and what the former are left to eat. Taylor also talks of the Latin American origins of the potato and what Irish people ate before its arrival. "There are so many ways to talk about the history of the Irish people and the history of the potato with just one potato," he says.

*"My Italian and Irish ancestors in this country were given many rewards for forgetting their peasant ways."*
*—Owen Taylor, founder of Truelove Seeds*

Culturally significant seeds can not only bridge communities and cultures through food, but also provide an opportunity for those from diverse origins to preserve their bonds to their roots and ancestral heritage by being able to cook the foods they grew up with and introduce them to their descendants.

East New York Farms! (see p. 184) and Soul Fire Farm (see p. 58) both grow seeds for Truelove Seeds. Callaloo seeds from East New York Farms! are some of the most popular available in Truelove's catalog "because of the number of Afro-Caribbean people or Caribbean people looking for their traditional foods," says Taylor. Soul Fire Farm meanwhile offers crops such as the Plate de Haiti tomato. "It dates back to before the Haitian revolution, so farm founder Leah Penniman is able to grow that seed and talk to the people who come to her farm about the Haitian revolution, which was the first and maybe only successful revolution where African people were able to overthrow the government and liberate themselves from slavery," says Taylor. "So it becomes an inspiring story that you can tell through a tomato that was there at that moment in history."

Collecting, growing, and sharing heirloom seeds contributes to protecting biodiversity and preserving cultural heritage for future generations. Growing culturally significant vegetables in cities can also become the spark that brings about more cohesive communities.

*Knowledge sharing is an important aspect of community farming. At East New York Farms! (see p. 184), manager Jeremy Teperman guides youth intern Nile while they harvest tomatoes and tomatillos (p. 43 top), and a list of crops for harvesting sits beside a wash station at the farm (p. 43 bottom), where crops include Sehsapsing blue corn and aji dulce peppers grown with seeds from Truelove Seeds.*

Harvest 7/24/20
Swiss Chard - 10
Kale - 10
Collards - 10
Beets - 15
Carrots - 20
Tomatoes - all ready
Scallions - 10
Malabar -
Bitter melon
Cucumbers
okra
Beans
Eggplant
all ready

# How to Grow in Small Spaces at Home

*Colorful containers and beautiful ceramic flowerpots packed with vegetables can bring some joy and breathe a new life into your space, while second-hand shops and seed exchanges can provide the tools you need on a budget*

**Be Imaginative with Space**

Not having access to a garden doesn't mean you cannot grow your own vegetables at home. Windowsills, balconies, fire escapes, and rooftops can make ideal growing spots. It's crucial to find a sunny growing zone since most vegetables need at least six hours of direct sunlight to thrive. Containers offer flexibility to be moved around. "Growing at small scale is often more about the joy of growing the plants than the quantity of food they produce. This is because when you grow your own food at home, you end up harvesting more than a meal—there is a fullness in the process. We learn to better understand our relationship with food by being responsible for its journey to our plate," says David Haisley, who tends his rooftop garden in New York City.

*Beekeeper Geraldine Simonis (above and previous page) moves her Thai basil plant onto her fire escape garden.*

*Outside her apartment window, Simonis has a small container garden where she grows kale, tomatoes, and other vegetables and herbs.*

*Urbanspace's former director of development, David Haisley, created a vibrant rooftop terrace garden outside his apartment in Brooklyn.*

*A family releasing ladybugs in a backyard vegetable garden in Brooklyn.*

*Zaro Bates, co-founder of Empress Green, an agriculture and social impact consultancy, grows microgreens on a windowsill in Staten Island.*

*Haisley harvests mint from his rooftop terrace garden in Brooklyn.*

*Haisley grows and harvests tomatoes and herbs.*

*The Brooklyn Grange Rooftop Farm's (see p. 226) Design/Build team built and maintains this container garden at a client's West Village townhouse in Manhattan.*

**Grow What You Like, with Wisdom**

Research what grows well in your area and consider vegetables that grow happily in a small container or pot, like strawberries, radishes, and herbs. Marigolds are a good idea to discourage pests and add a splash of color, and a plus is that honeybees love them. "You need to be realistic with the space you have. At first I had a list of 30 crops I wanted to grow but ended up choosing 4 and building from there," says Geraldine Simonis, who grows her own on her fire escape.

**Consider the Seasons**

If you participate in seed exchanges or you saved them from last year's harvest, it is recommended to check for the most suitable time to sow seeds. Alternatively, store-bought seed packets usually contain information about when to sow them. According to Simonis, buying local seedlings is advisable since they are likely to grow better in your area.

Each space is unique and the most critical step is to start, learning what works better for your space along the way.

# KAT LAVERS

*A food gardener who has designed a permaculture kitchen garden in her Melbourne backyard*

Cultivating food is in Kat Lavers's genes. Having grown up in a food-gardening family, she began to produce her own food while studying environmental science and human geography at university. "I was desperate for some practical skills," she says. "I was doing really theoretical study and I felt like my hands were being neglected." As a student spending a considerable amount of her budget on store-bought fresh produce that was nowhere near as interesting as the foods she had grown up with, she decided to start growing her own. "I made many mistakes but still got enough food to motivate me to keep going," she says.

Her northern Melbourne kitchen garden, The Plummery, is conceived as an urban permaculture system that mimics patterns found in nature. She grows a myriad of fresh produce, from herbs and berries to tomatoes, and also has multiple fruit trees, including an almond tree propagated from one located in the garden of her childhood home. Lavers started her ever-evolving garden around 2009, soon after she bought her property. In 2020, she collected 450 kilograms (990 pounds) of fresh produce from her 100-square-meter (1,100-square-foot) growing area—enough for two adults to have most of their annual fresh produce needs covered, with the exception of bulk crops such as potatoes. She estimates maintenance work takes an average of four hours a week.

*"I was desperate for some practical skills. I was doing really theoretical study and I felt like my hands were being neglected."*

Lavers's day job is in local government, running a sustainable gardening education program as well as supporting urban agriculture projects. She is aware that yields are the reward that motivates urban food gardeners to keep going. "For us, the main barrier was not having the skills and knowledge. I spent a couple of years messing around, seeing what happened. Loads of things I tried failed before I found a better way of doing them. What's driven me to teach is that I really want to help people shortcut and get to something that's successful more quickly, because that yield of food is what motivates people to keep going—if they don't experience that, most people would give up."

Her garden is informed by the permaculture process, which is based on three ethics: care of earth, care of people, and fair share. The garden's design has evolved over the years in response to the diverse challenges Lavers has faced. She now maintains raised beds instead of growing on the soil directly, due to the soil's lead contamination; conversely, this is not a problem for fruit trees since the lead in soil transfers mostly to the trees' leaves and roots, rather than the fruits. Integrating quails with a compost system called "deep litter" forms part of her balanced approach: "We receive compost, eggs, and pest management from them, and they have somewhere that they love to scratch around to stay mentally and physically healthy," she says.

*"Cities are never going to be self-sufficient, but by growing food at home, we can contribute to the solution."*

Growing her own has added new layers of significance to Lavers's approach to food, motivating her to support local producers and view as a resource what before was a waste (for instance, she now feeds her soil with food scraps). "Now I have such a respect for food, and all the effort, the water, the nutrients, the soil that it takes to grow it. And that translates into other behaviors, like trying to avoid food waste, which is a huge problem in our food system," she says. "I also have massive respect for farmers. Farming is really difficult. It's getting harder with climate change in many parts of the world, and it's incredibly challenging to make it work economically, especially when you are doing small-scale regenerative farming. I don't think I'd have realized that had it not been for the experience of growing food myself."

Food gardening can contribute to building resilience in urban environments, from supporting biodiversity by creating habitats to helping combat the urban heat island effect. "We seem to be facing so many different types of emergency: climate, humanitarian, biodiversity, health ..." Lavers says. "Cities are never going to be self-sufficient, but they can have more of a buffer. So, by growing food at home, we can contribute to the solution; we can make things a bit more resilient, more comfortable when we get challenged by these sorts of events."

*1] Gardener Kat Lavers holds a homegrown bounty in her edible garden (p. 49)*
*2] Kat with one of her Japanese quails* (Coturnix japonica)*(p. 50 top)*
*3] Japanese quails can be an alternative to backyard chickens in urban areas, as they need less space and make less noise (p. 50 bottom)*
*4] Lavers's home garden, where she grows an assorted variety of crops including tomatoes, beans, and herbs (p. 51)*
*5] Laver's home garden includes places to relax and enjoy the space, which encourages observation and learning (p. 52 top)*
*6] Eggs from the Japanese quails: each hen lays eggs with a distinctive shape and color (p. 52 bottom left)*
*7] Lavers converts part of her bounty into homemade sauces, pickles, and preserves (p. 52 bottom right)*

# Harvesting our Sovereignty

*How Black, Brown, and immigrant women birthed the urban farming movement*

BY FARM MANAGER CHRIS BOLDEN-NEWSOME

The story of urban farming in the U. S. today is complex. Much of the official narrative ignores or sidelines the Black, Brown, Red, and Yellow people—and especially the women—who have since slavery kept what the authors of *In the Shadow of Slavery* call "botanical gardens of the dispossessed." For many people of color who raise urban gardens, a common theme is escape, be it from economic pressures or from racist White power structures.

The urban farmers usually covered by the media are young White folks who left university degrees and good jobs to work in poor communities teaching kids of color to garden.

However, Karen Washington, a mother of the food sovereignty work of New York City, traces urban agriculture history to the "victory gardens" of the two world wars. During the war years, the U. S. government encouraged citizens, largely women left behind, to safeguard the economy by growing as much food as they could. If the marketing in those days showed smiling White women, the reality according to Washington was that poor African American and immigrant women in cities and towns were already growing gardens. In fact, she teaches, "Black and immigrant mothers created many of the features of city gardening from necessity; container gardens, raised beds, rain barrels, bio-intensive," and other now common techniques. In this way, the wartime generations did not starve and they depended on cultural dishes for both comfort and survival.

Though urban farming has become a hip lifestyle choice for many young White professionals, it remains a matter of cultural, spiritual, and nutritional sovereignty for people of color and immigrant city farmers. Hailing from the Carolinas, the Virginias, and the Caribbean, Philadelphia's population is majority African American, with many Black families who are only a couple generations removed from the South. After WWII, Black people's land and property was lost through discriminatory policies, and practices such as "redlined" communities, where legislation prevented Black people from buying property, and the denial of GI Bills to Black veterans, which resulted in a transference of wealth to White Americans.

Despite years of disenfranchisement in the city, due to White flight, government disinvestment, and hyper-policing, African influences in culture persist. For some Black Philadelphians, knowing their family origins outside the city is a crucial link to a story beyond the statistics of poverty, hunger, disease, and addiction. Today many Black Philadelphians still proudly name the villages from whence their families came.

At Sankofa Community Farm in southwest Philadelphia, we work to recover the tools of Black and Brown food sovereignty, and reconnect Black Philadelphians physically and intellectually with their foodways and other ancestral healing tools. We cultivate

traditional crops, pray together, keep seeds, cook our foods, and tell stories that center our experience. In a unique partnership on the land of historic Bartram's botanical garden, the work of food sovereignty has aided Philadelphians of color to connect across centuries and miles to the movements of our ancestors through the rituals of farm work. The act of cleaning seeds with traditional songs becomes a sacramental moment, an action that embodies ancestors and stories lost to us by time and struggle. Sankofa Farm calls Black and Brown people to examine their identity in community, beyond the dominant narrative of rugged individualism, to consider how folks of color might show love in practical ways, such as greeting neighbors on the street and sharing our crops. Seeing my people take their healing in hand, I imagine the tall stalks of okra in the garden of an ancestress whose face I have never seen and I thank her.

*Originally from the Mississippi Delta, Chris Bolden-Newsome is farm co-director, crop management teacher, and fourth-generation "free" farmer at Sankofa Community Farm, where he follows natural agriculture practices focusing on healthy soils.*

# Food Sovereignty and What We Can Learn from Cuba

*Urban agriculture can increase access to healthy, locally grown produce if approached from an equity lens*

The global peasant movement La Vía Campesina coined the term "food sovereignty" in 1996. Unlike food security, which in its basic form means to have access to sufficient, nutritious food, food sovereignty also covers people's right to define agricultural strategies, including access to land, water, and other resources, prioritizing local, ecologically appropriate production, fair conditions for agricultural workers, and affordable access to seeds.

The world's largest organic farming experiment took place in Cuba during the early 1990s. In the 1980s, the island's economy had hinged upon trade with the Soviet Union. The collapse of the latter meant the end of high-input agriculture for Cuba. The island was forced to transition almost overnight from an industrial food system that relied heavily on agrochemicals and imports to a self-sustaining, agroecological, local approach to food production. During the "Special Period" that followed, thousands of popular urban gardens emerged out of necessity. The grassroots response to the food shortages brought about by the loss of trade with the Soviet Union soon received formal government support. Today, the country is widely considered a pioneer in urban agriculture. As stated in research led by professor Miguel Altieri, 50 percent of the fresh food on the island is produced in urban and peri-urban agricultural settings.

Seed saving is crucial for food sovereignty and central in the work of agronomist Dr. Humberto Ríos Labrada, a Goldman Environmental Prize winner for his contribution to agricultural transformation in Cuba. "When the crisis came to Cuba, one of the sectors that collapsed first was the seeds," he says. "I traveled around the country collecting seeds mostly from small farmers who had been discriminated by the Green Revolution model."

Labrada considers the participative dissemination of seeds his biggest contribution to the drastic reform in the country's food system. "For the first time in the history of Cuba, I asked small farmers, rural and peri-urban, which seeds they wanted to select. Later, urban farmers followed, " he says. "We started with around 25 farmers, and in a few years, we had a network of 50,000." Each small farmer who took part in the first seed fair then went on to organize similar fairs, promoting seed saving, seed sharing, and local agricultural production. "In Havana, like any other big city, there are people from many places, and their criteria to select seeds are even more diverse. People select different seeds depending on their culture. It's a beautiful way of trying to increase biodiversity," he explains.

Vivero Alamar is a successful *organopónico* (urban garden) located in the Havana suburbs. Run as a cooperative with over 150 members, it stands in a high-density area, offering the community easy access to affordable, fresh produce. Fruit and vegetables are grown following an intensive organic farming model. The garden is run as a circular ecosystem where everything from manure to compost is managed in-house. Another example of urban agriculture in the capital is La Sazón, an *organopónico* that is sandwiched between buildings in a high-density area. As with all *organopónicos,* La Sazón is government regulated, selling its organic produce directly to consumers at low prices. By encouraging the participation of residents and empowering communities, the urban agriculture movement in Cuba gained social acceptance; it created jobs, offered affordable, organic produce, and beautified neglected areas.

*Cuba was forced to transition almost overnight from an industrial food system to a local, agroecological model.*

Leah Penniman, the founding co-director of Soul Fire Farm and author of the book *Farming While Black*, works toward ending racism and injustice in the food system. Food deserts—areas with a lack of buyable fresh produce—are a visible symbol of food injustice in the United States. Penniman considers the situation food apartheid, a term coined by activist Karen Washington. "It's not a natural desert, it's a human-created system of segregation," she says. "You can trace it back to a history

*Cuba has become a leading force in organic urban farming, with hundreds of urban gardens known as* organopónicos. *Streetscape and overview of* organopónico *La Sazón in Havana (p. 56), sandwiched between buildings in a high-density area (p. 59 top) where local residents grow food (p. 57). Volunteer hands out free produce at ENYF!'s (see p. 184) weekly food distribution at their Pink Houses Community Farm (top); Leah Penniman of Soul Fire Farm runs programs that address food sovereignty issues affecting underserved communities (p. 59 bottom).*

of redlining housing discrimination and restrictive zoning, as to why folks of color predominantly live in certain neighborhoods, and these neighborhoods have a scarcity of farmers' markets, grocery stores, fresh food outlets, and also have low income and no wealth. It's that history of systemic racism that has resulted in this ghetto-ization of food."

> *"A food desert is not a natural desert, it is a human-created system of segregation, it is food apartheid."*
> *—Leah Penniman, founding co-director of Soul Fire Farm*

From her Upstate New York farm, Penniman runs a number of programs that address vital food sovereignty issues affecting underserved communities, providing families with the tools and guidance to grow their own produce. "One of these programs is Soul Fire in the City, which builds raised bed gardens for survivors of food apartheid in our local area of Albany and Troy," she says. "There are just under 50 families in the program. Each family gets the lumber and hardware. The raised-bed is built by our volunteers, set with compost and topsoil. We bring the seeds and plants, and have a number of classes, as well as a phone-a-farmer hotline, so they can call and get advice about their garden. It's been beautiful because especially during the pandemic, we heard from families that this is not only a crucial source of food, but also a source of connection for their children to the earth, a family activity that is helpful and hopeful in times of despair, so that's been very meaningful."

Improving access to local, healthy food for underprivileged communities is part of East New York Farms!'s mission too. "At our Pink Houses farm, all the produce grown gets distributed to the community for free," explains Iyeshima Harris, the project's director. The community farm, located in the Pink Houses public housing complex, grows around 21,000 kilograms (46,000 pounds) of produce annually. Volunteers, youth program interns, and Pink Houses residents help maintain the space.

Growing food in urban settings can take many forms, from an individual act to a collective effort. It may not be realistic to feed entire cities on locally grown food, but urban agriculture can be a significant part of the solution to achieve fairer levels of food access, especially if approached from an equity lens.

Organopónicos *usually have a market stall on site, where affordable produce is sold within steps of where it is grown, like the one at* organopónico *El Minero in Santa Clara (p. 61 top right), or El Japonés in Havana's Vedado district (p. 61 bottom right). The stalls are not only a selling point, but also a community gathering space.*

ORGANOPONICO El Minero
EXCELENCIA NACIONAL
EL MINER

# Aquaponics, a Sustainable Vision of Tomorrow's Cities

*Oko Farms is the largest outdoor aquaponics farm in New York City, and the only one with public access*

Co-founded by Yemi Amu and Jonathan Boe in 2013 and located in a formerly vacant lot in East Williamsburg, Brooklyn, Oko Farms has evolved through the years into a production, research, and educational farm.

"We have two missions," says Amu, the farm's director. "The first is to practice and promote aquaponics as a sustainable farming method that can increase food security and protect us against the impact of climate change. The second is to spread the knowledge required for people, regardless of race or socioeconomic background, to be able to practice aquaponics for themselves."

Aquaponics brings together aquaculture and hydroponics. "It is a way of producing food in a symbiotic ecosystem, where you are raising freshwater fish, and recycling the waste water from those fish to grow vegetables, and in the process, the roots of those plants are also helping filter the water for the fish," explains Amu. It's a scalable model that can work from individual to large commercial operations, and because it is a closed system that recirculates water, it uses significantly less amounts of it than conventional agriculture.

Following her master's degree in health and nutrition education at Columbia University, Amu began educating people in how to eat a healthy diet. Her hands-on approach included trips to the farmers' market and advice on eating seasonally. "Once you know how to buy fresh produce and how to cook it, you should probably know how to grow it too. What I am doing now is completing that cycle of educating people from seeds to plate."

Education is an integral part of Oko Farm's mission. "The major way that we impact food justice is by giving people the tools to grow their own food for themselves," Amu says. "It is more powerful and empowering for you to have the knowledge to grow food, than for you to be dependent on people to give you food." She promotes local access to food tirelessly from her farm, and for almost two decades has contributed to founding dozens of edible spaces at schools and community organizations across New York City to make fresh produce more accessible. "If we had more farms in our community, then more people would have access to food, and there would be less issues with food insecurity," she says. Pre-Covid-19 figures suggested that 12.9 percent of New Yorkers were food insecure, with numbers more than doubling since the pandemic began.

Oko Farms

At the beginning, Oko Farms only grew basil, a crop they were selling entirely to a nearby pizzeria. When the restaurant halted their orders overnight, they took it as a chance to reroute and really do what they had intended. "It was both a shock and a disappointment, but also a great opportunity," says Amu. "We are a small farm, and in order for us to provide for our customer at the scale they wanted, we were forced to grow only one thing. And that's not a good way of educating people. For many of our visitors seeing an aquaponics farm for the first time, it was like, 'OK, so with aquaponics you can only grow basil.'" Researching what was possible to grow in an aquaponics system beyond basil followed that initial hurdle, and today the farm grows a wide range of different vegetables.

The most important aspect for Amu in determining what to produce is to ensure it can survive in a city climate and be grown in aquaponics. The farm has three seasons—spring, summer, and autumn—and they pick their crops accordingly, while their fish live through all seasons, including winter. "We also try to think of growing produce that speaks to a diversity of cultures," she says. They grow jute (*ewédú*), a fibrous leafy vegetable very popular in Nigeria, where Amu grew up. "It's something people come to us for, because you can find jute frozen or dried, but not fresh at the market." Another example is lemongrass leaves, one of the most popular vegetables they sell.

*"We have a mission to spread the knowledge required for people, regardless of race or socioeconomic background, to be able to practice aquaponics for themselves."*

At the same time, Amu never takes off her educator hat: "I also think, 'How can we use these vegetables to teach history, climate change, food justice, and how do we use them to provide a fun cooking experience for people?'" She has found that sorghum, an ancient grain, is ideal for teaching. "It's very easy to grow in soil or aquaponics; it's one of the oldest farm crops in the world," she says. "Agriculture started in Africa, and sorghum is one of those grains that was grown for food security because it can survive in floods and

*1] Yemi Amu, founder and director of Oko Farms (p. 63)*
*2] Poles are used to move the floating rafts in order to reach and harvest the produce (top)*
*3] Volunteer Vianca harvests onions for the Saturday farmers' market at Maria Hernandez Park (p. 65 top)*
*4] Plants on the aquaponics farm are grown inside floating boards with their root systems in the water (p. 65 bottom)*

*5] A floating board is pulled from the water to harvest lemongrass, exposing the root structure beneath it (p. 66 top)*
*6] The tank containing the farm's multicolored koi fish is dug into the ground for insulation without heat or cooling (p. 66 bottom left)*
*7] Amu harvests jute, also known as* ewédú*, a popular crop in East, West, and North Africa, and the Middle East (p. 66 bottom right)*
*8] Amu harvests produce for the farmers' market. The large tent behind her houses a tank containing koi fish (top)*

droughts." Ornamental and edible fish are raised for their key role in aquaponics, but they are also used for culinary workshops, and for teaching people how to eat fish from nose to tail, and how to clean and gut them.

> *"Once you know how to buy fresh produce and how to cook it, you should probably know how to grow it too."*

The farm closes during the winter months, when Amu focuses on her educational work in schools. Since the Covid-19 pandemic, they have begun providing remote learning opportunities for school groups and adults. Having different income streams is vital for any farmer, and even more so in a city, where space is limited and urban farmers can't grow at the scale rural farms do. So in addition to selling vegetables and providing workshops and tours, Oko Farms offers design advice and technical aquaponics training guidance. "When you are farming, you are at the mercy of nature in a lot of ways, so having different sources of income is a way of protecting your farm," Amu says.

The farm was built as a collaboration between its co-founders Amu and Boe, the community gardening program GreenThumb, and Brooklyn Economic Development Corporation. Green spaces like Oko Farms, located on city-owned land, have recently been subjected to new restrictions introduced by the Parks Department, which has made it unviable for some of them to operate. Oko Farms is now in the process of moving to a new 1,000-square-meter (10,800-square-foot) location at the Weeksville Heritage Center in Brooklyn, a space that educates people about the history of Weeksville, the first intentional Black community in New York. "This center is not only making people aware of this historic community, but also creating an opportunity for us to continue that legacy of self-determination and empowerment for African Americans," says Amu. She also plans to keep the East Williamsburg farm available to the community. But wherever the location, from her farm to the edible gardens she helped found, Amu will continue to work relentlessly for a more just, sustainable food system.

*9] Amu harvests rice, which will be hung to dry before being hulled (p. 68 top)*

*10] Crops grown at Oko Farms are used in cooking demonstrations, which are key at the farm. This dish contains stewed okra with fermented locust beans, smoked-dried catfish and chili with a garnish of scent leaf (p. 68 bottom left)*

*11] Harvested lemongrass is bundled and tied for the weekend farmers' market (p. 68 bottom right)*

*12] Amu ties bunches of mint for a local farm box client (top)*

# Education and Youth

*Growing up in contact with nature can be beneficial for both the planet and those who will look after it in the future*

An appreciation for nature comes effortlessly to those who have been exposed to the natural world from an early age. Today, nonprofits and progressive organizations are pushing forward a structural change, expanding the options for city kids of any socioeconomic background to engage with nature and learn to respect it through their programs.

With a goal of transforming children's health with hands-on educational experiences, acclaimed chef and activist Alice Waters founded the Edible Schoolyard Project in Berkeley, California, in 1995. "I believe that all children have to be connected to nature at a very young age, because nature is our mother, our teacher, and it's important to understand that's where our food comes from," she says. In the Edible Schoolyard's kitchen and garden classrooms, kids are exposed to nature and food through the academic curriculum, from history to languages. "When people think of childhood, they remember what something tasted like, some place when their senses were engaged," explains Waters. "I think that's what it's all about—opening up your mind to your senses. Food is all about those senses. They are the pass-ways into our minds, so the earlier children experience that connection, the more likely they'll feel empowered around eating for the rest of their lives. I call it a 'delicious revolution' because it's not difficult to do. The older you get, the more indoctrinated you are by our fast-food culture, so we need to reach children when they are young and open and bring them into that relationship."

In New York City, where green space is precious, nonprofit organization City Growers uses urban agriculture as a vehicle to teach kids and youth about where food comes from and why it matters, touching on subjects from environmental literacy to food

education. "Urban agriculture spaces like Brooklyn Grange are our living, learning laboratories," says City Growers' executive director Becca Di Meo. "We use non-traditional spaces in the urban environment, such as rooftop farms and school gardens, to engage youth in experiential learning that nurtures a life-long relationship to food, their health, and the natural world."

*"Nature is our mother, our teacher, and it's important to understand that's where our food comes from."*
*—Alice Waters, chef and activist*

City Growers' programs and workshops have a tiered pricing structure that helps reduce barriers to access, offering lower fees to schools with less resources. According to program manager Emma Taliaferro, after attending the workshops young people start to recognize the city as a place where nature can thrive. "We highlight the importance of urban green space like parks, community gardens, and even street trees—for example, one linden tree can provide honeybees with enough pollen to produce 18 kilograms (40 pounds) of honey—and that even the smallest creatures, like bees and worms, have a special place in our city's ecosystem."

Their original curriculum teaches science standards but also incorporates hands-on activities that range from farm tours to compost workshops, seed saving, harvesting, and growing. After-school and summer-camp programs offer the chance to plant seeds and tend them as they grow, and include peer-to-peer teaching models for both cooking and gardening. A change in attitude is often perceived after only one workshop. "There are insects flying and crawling all over the farm," says Courtney Epton, City Growers' director of education. "Upon first arriving, a young person flinches and cringes every time an insect gets close. In just 90 minutes, after holding a worm and learning about compost, after checking out the inside of a beehive, this same young

*More and more initiatives are being created to bring city kids closer to agriculture. City Growers' activities in New York include Farm Explore workshops at Brooklyn Grange Rooftop Farm (see p. 226), demonstrating how to distinguish crops from weeds (pp. 70 and 71), and how to plant seeds that future students will harvest. In Chicago, Windy City Harvest offers teenagers an internship program. Its North Lawndale location (p. 73 top) is nestled between the El tracks and Ogden Avenue at the beginning of Route 66. Staff members at the McCormick Place Rooftop Farm (p. 73 bottom) harvest produce that goes directly to the kitchen of Savor restaurant below.*

Windy City Harvest
Windy City Harvest
Windy City Harvest

person is watching a bee fly from flower to flower mere inches from their nose."

Academia is a career path that students pursuing graduate degrees might follow. At AgroParisTech, a Parisian university that specializes in technical agronomy studies, theory and practice are integrated. The institution houses an experimental 860-square-meter (9,260-square-foot) rooftop farm, designed and built in partnership with Topager, where students can participate in hands-on research through internships.

Using urban by-products to create sustainable soils for rooftop food production is one of many projects AgroParisTech has developed. The university's study of these man-made soils, known as Technosoils, spanned five growing seasons, focusing on the value of food production and the ecosystem services provided to the city by rooftop gardens. "Our experiment mostly aims to conceive and test substrates made by urban residues, and to participate in urban metabolism through urban agriculture on roofs, or over urban soils when they're not cultivable because of contamination," explains professor Christine Aubry. The materials used to create these innovative substrates include woodchips, green compost, mushroom substrate, compost from urban farms, and other urban organic residues.

Windy City Harvest is an urban agriculture education and jobs-training initiative at Chicago Botanic Garden. It offers programs with a focus on sustainable urban agriculture and food systems, from managing a beehive to selling food grown on farms at local farm stands and markets. The Youth Farm program, which runs from March to October, offers teen farmers paid, on-the-job training. "Youth farmers are coming to us from a lot of different backgrounds, and we really focus on how plants can help us connect to social-emotional learning, which can make us successful in whatever path we choose," says Eliza Fournier, the program's director.

*"We focus on how plants can help us connect to social-emotional learning, which can make us successful in whatever path we choose."*
*—Eliza Fournier, urban youth programs director of Windy City Harvest*

The main mission of Windy City Harvest is to bring food, health, and jobs to the communities where the farms are located. They have 16 Chicago-area sites covering 3.6 hectares, and youth farmers have the option to join the program for one or multiple years. On the farms, youth farmers grow culturally appropriate, high-producing crops and can take home any produce they want. "Covid-19 has underscored the disparities between different communities in Chicago, and the disconnect between people who live in different neighborhoods," says Fournier. "Growing plants and food on an urban farm can be healing, and when people feel helpless and hopeless, it's a way to gain a lot of hope and joy." Youth can benefit deeply from an environmental education, acquiring valuable knowledge and soft skills to equip them for the future.

*At AgroParisTech in France, students tend to the rooftop garden designed by Topager (see p. 22). A student removes dried plants to make room for new planting (p. 75 top), different types of urban residues are used to create Technosoils, which give a second life to local waste produced by the city (p. 75 bottom left), and farm manager Charlène plants seeds for sweet peas (p. 75 bottom right).*

MARC DE
BOIS

# MEREDITH HILL

*An educator with a focus on sustainability and nature who promotes the power of youth voices in education*

An accomplished school designer, garden educator, and school leader, Meredith Hill is currently principal of MS 371, School of Earth, Exploration, and Discovery (SEED: Harlem), a new middle school launched in 2020 in New York City. She believes in the power of education to connect students to nature and encourages learning experiences in which pupils play a meaningful role.

Originally from Haverhill, Massachusetts, Hill moved to New York City to attend Barnard College, a women's liberal arts college. She majored in theater directing, minored in dance, and completed the Barnard Education Program. Her teaching career began in 2007, as a founding faculty member at Columbia Secondary School for Math, Science, and Engineering, a public school partnered with Columbia University, where she had previously completed two master's degrees at Teachers College.

She stayed in the job for over a decade, during which she not only taught sixth-grade English, but also launched various inventive projects, from a youth summer program with a focus on environmental activism and engineering to a student magazine and a garden compost system that at its peak processed 75 percent of the cafeteria's organic waste.

Hill followed this by founding another remarkable project: the Columbia Secondary School Community Garden, located on a formerly vacant Parks Department lot. "I designed and taught courses ranging from garden-to-table cooking to high school agroecology, and hosted weekly open-garden hours to build our garden from the ground up," she says. "The goal was to educate about sustainable agricultural systems."

*"Students need to see voices of power that reflect themselves; we need to help our future leaders see their potential and the power in their own histories."*

Today, as principal of SEED, her focus is on social and environmental justice and hands-on, project-based learning. "Working toward equity is a critical obligation of educational institutions," she says. "We believe in the power of youth voices and the need for education to yield authentic outcomes for them. We aim to rethink

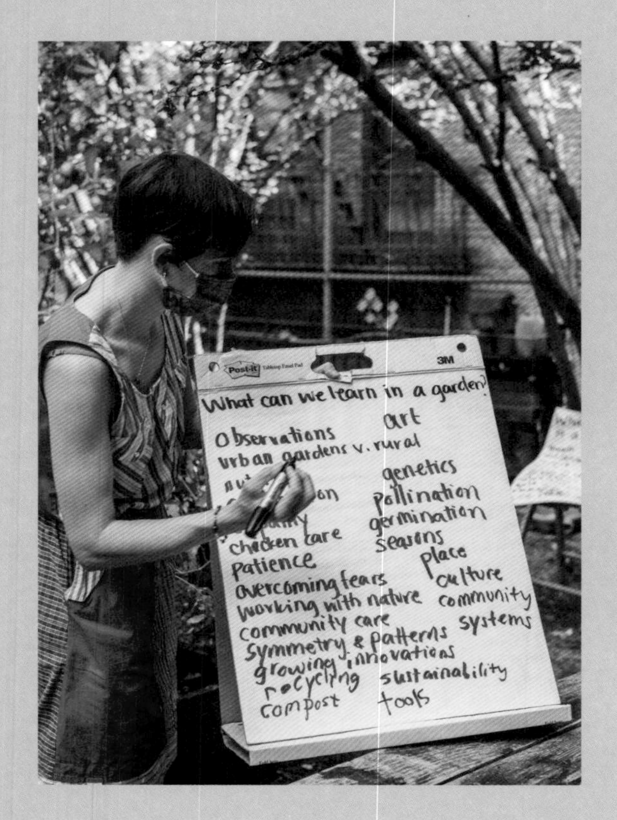
Post-it
3M
What can we learn in a garden?
Observations
art
urban gardens v. rural
genetics
pollination
germination
chicken care
seasons
patience
place
overcoming fears
culture
working with nature
community
community care
systems
symmetry & patterns
growing innovations
recycling
sustainability
compost
tools

THE
HOT CHICKS
ROOM

social studies so that students see voices of power that reflect themselves; we can only change oppressive systems by challenging dominant narratives of power and helping our future leaders see their potential and the power in their own histories."

Hill sees her role as a teacher-facilitator; she believes in allowing her students to participate actively in their own learning process, and places the environment at the heart of the curriculum. "Our planet is in crisis, and in order to make a true, lasting shift, we need to raise future leaders who care about the environment and who see their power as part of an ecological system," she says. "We need to not only teach our youth to be tech savvy and smart, but to care, to connect with the innate human relationship with the Earth."

*"Our planet is in crisis, and in order to make a true, lasting shift, we need to raise future leaders who care about the environment and who see their power as part of an ecological system."*

Before the school opened, Hill interviewed local students and families to hear their thoughts on education, and SEED's design team used collaborative protocols and included teachers, students, and community members. They aim to build partnerships with environmental groups such as Earth Matter and the youth organization The Brotherhood-Sister Sol in the near future. Establishing a school in the midst of a pandemic was challenging and has required some alterations to their plans: "It reminds me to be humbled by our humanity and to embrace the value of learning from the world around us," she says. "After all, isn't that the most transformative education that we aim to build for our students?"

According to Hill, people are starting to see the need for this progressive approach to education, a system not unheard of in more affluent communities but a rarity in the public school system. "My dream is to think about this not just as a school, but the start of a movement to change what's possible in education," she says. "Currently we are building a picture of an education that is so much more holistic and really values deep learning. Systemic change takes showing what's possible and then sharing it."

*1] Meredith Hill, educator and principal of MS 371, School of Earth, Exploration, and Discovery (SEED: Harlem) (p. 77)*
*2] Marisa DeDominicis, director of Earth Matter, leads students at the Compost Learning Center (p. 78)*
*3] Hill helping students prepare for a presentation on raising urban chickens (p. 79 top)*
*4] Former students now hold leadership roles at the Columbia Secondary School Community Garden in Harlem (p. 79 bottom)*
*5] Hill introducing teachers to garden-based education (p. 80 top left)*
*6] A former student shares a poem about the chicken he raised as a class project (p. 80 top right)*
*7] Composting at Earth Matter. Chickens help to aerate the piles and provide manure (p. 80 bottom left)*
*8] A former student holds a chicken she raised (p. 80 bottom right)*

# Creativity and Food Gardening in One Open, Welcoming Space

*PAKT is a multidisciplinary space in Antwerp, Belgium, where urban farming co-exists with over 20 modern businesses*

Located in the heart of Antwerp, Belgium's second city, PAKT was launched in 2017 by Stefan Bostoen and brothers Ismail and Yusuf Yaman. The trio have transformed what was once a bunch of derelict warehouses into a vibrant space with dozens of thriving small businesses and a productive urban rooftop farm. "Before the start of the renovations, various artists and musicians were invited to use the old warehouses," explains Yusuf Yaman. "They inspired me to create PAKT, a lively hub for creative and sustainable entrepreneurs." Today, different businesses co-exist there, including environmentally friendly restaurants, PR and content agencies, and a cooperative that buys agricultural land and rents it to organic farmers through career-long contracts.

The foundations of PAKT's 1,800-square-meter (19,400-square-foot) urban rooftop farm were established in 2016, when agrobiotechnologist Bram Stessel and community expert Adje Van Oekelen were hired to work on the design of the new urban agriculture project. "Besides being an organic farmer, Bram also played in a band that rehearsed on the site," says Yaman. "As I'm a gardening enthusiast myself, together with Bram, we conceived the idea of growing food on the roofs."

*"In my eyes, urban agriculture mainly ensures that city people discover a connection with their food and surroundings."*

At the time, Stessel had just started a job as a postal worker after a bittersweet farming experience in Belgium, where he says access to farming land is unaffordable for the majority. His new career lasted barely a week. "I got a phone call from Yusuf, who wanted to test if vegetables could be grown at the old warehouse," remembers Stessel. "He hired me to do a whole year of research to test cultivation techniques and water collection. This was the most intense year of my life. Yusuf wanted to develop a green site for creative companies, and a roof garden was a must for him. After a year of many failed tests, we finally found the right ways to collect water and grow vegetables on a roof, and PAKT was born. In the winter of 2017, I started the construction of the roof farm with some friends, and set up a cooperative with Adje to run the roof garden."

PAKT

By integrating urban agriculture in a functioning, modern space, PAKT offers a chance for creative businesses to develop in the city but be closer to nature. "In my eyes, urban agriculture mainly ensures that city people discover a connection with their food and surroundings," says Stessel. "We don't grow vegetables; we grow new communities of citizens. By taking care of your environment, your environment will give you back healthy food."

*1] PAKT is a unique collaboration of creative entrepreneurs and urban farmers in Antwerp, Belgium (p. 83)*
*2] Membership includes the option to bring up to five friends onto the rooftop, and every year PAKT welcomes thousands of visitors (top)*
*3] More than 90 different perennials grow on PAKT's roofs (p. 85 top)*
*4] Each month, two kids from the local school attend a free workshop organized by volunteer farmers (p. 85 bottom)*

Stessel soon realized that growing crops on a roof is completely different from growing them at ground level, but his year-long experiments allowed him to become familiar with a new way of working. Maintaining soil life and preventing it from turning into dead substrate was one of his biggest challenges, but he found water collection and circulation was easier to control on the roof than on the ground. "We try to look at nature as much as possible and apply it to the ecosystem of the roof," he says. "Our basic philosophy is to constantly stimulate soil life to grow plants as healthily as possible. For us, the hay bales in which we grow the vegetables are the motor that starts the entire transformation process, from organic matter to a living, fertile soil. After the bales have been used to grow many of our vegetables for one season, we apply them as mulch on our substrate, and in doing so, they also bring life back into our soil. Without the hay bales, we wouldn't have a soil life on the roof."

*"We don't grow vegetables; we grow new communities of citizens."*

The urban farm is run as a cooperative: the farmers and a small group of investors have shares. Members pay a monthly €50 ($60) fee, which gives them access to the roof garden. Up there, they can harvest as many vegetables as they want, and also use the space to relax. PAKT believes in transferring as much knowledge as possible, so workshops covering how to grow food are available twice a month for members. "The lessons on growing vegetables are taught by farmers like me," says Stessel. "We offer a platform for young farmers to earn from teaching on our roof garden, and they can also sell vegetables from their

db

When life gives you lemons make
STEP 1. PICK A JAR
STEP 2. ICE ICE BABY
STEP 3. PICK A GIN

*5] Some members join to learn about gardening, others for the social activities and community (p. 86 top)*
*6] A birthday celebration for one of the rooftop farmers (p. 86 bottom left)*
*7] Pizzaiolo from Standard pizzeria during a kids' workshop at PAKT's birthday weekend (p. 86 bottom right)*
*8] The sunny rooftops at PAKT are a popular gathering place (top)*

PAKT 'n aperitief
DAKbloemen en kruiden
Groenteweelde
Courgette
Bietjes
Bloemkool
Boontjes
Tomatenchutney
Pickles
Dosa's
Garnituur
DAKbloemen
Kraakverse groenten
stadsboeren van PAKT
Pittige pekels
De Pekelcomp
Gecreëerd en geserveerd met liefde
Chantal van
GRØNT

aubergine
paprika
€5/kg
courgette
€1

farms outside Antwerp to restaurants through our online shop."

*"We offer a platform for young farmers to earn from teaching on our roof garden, and also sell their vegetables."*

Members who have gained enough knowledge can opt to pay half the monthly fee in exchange for providing guidance to fellow rooftop farmers on the various maintenance tasks needed. To make the space more inclusive, all members are allowed to share their membership with one adult and bring five non-paying guests to the roof every day. There are about 75 paying members and double the amount of rooftop farmers. According to co-founder Adje Van Oekelen, "We have had thousands of guests each year. Our aim is to inspire as many people as possible."

PAKT's path toward a profitable urban farming model includes consulting for a range of companies, from architects to developers, offering their knowledge in establishing urban green spaces. "Our vision is to develop the local food web as much as possible by inspiring and stimulating both professionals and citizens," explains Stessel.

*"Our vision is to develop the local food web by inspiring and stimulating both professionals and citizens."*

Both his and Oekelen's office is located on the rooftop farm, an ecosystem in which three fruit trees exist alongside over 100 varieties of berry bushes, vegetables, herbs, and mushrooms, as well as chickens and edible fish. Members do most of the everyday tasks like weeding and watering, but Stessel ensures everything works

smoothly. "Our roof garden is constantly open and our members come and go whenever they want," he says. "There's always somebody there who has questions, about gardening or about life; if so, we solve them while mulching or shoveling compost together."

*9] Local farmers use PAKT as a platform to sell their organic vegetables. There is an online shop for Antwerp-based chefs and caterers, and a weekly farmers' market (p. 88 top right and bottom)*

*10] PAKT offers the chance to gather and meet other members of the community on the rooftop, especially during the warmer months (top), and at dinners (p. 88 top left)*

# How to Use Natural Dyes at Home

*Dyeing your own fabrics naturally is a millennia-old process that allows you to experiment with different tones and patterns*

Different techniques can be applied in natural dyeing, and essential tools include the ingredients used for dyeing, fabrics, a stainless-steel stock pot, and rubber gloves.

Liz Spencer is a natural dye artist who shares her knowledge on coaxing color from plants through her workshops.

Bundle dyeing is an intuitive, water-saving method that produces a unique pattern every time. Before starting the process, fabrics need to be pre-treated with a mordant solution that acts as a glue that sets dyes on fabrics, preventing colors from fading out. Alum, a metal salt that occurs naturally, is commonly used.

Spencer uses dye flowers like chamomile, hibiscus, or marigolds as dyeing materials. These are sprinkled on one half of the

*Jars of organic U. S.-grown indigo pigment with a jar of fresh Brooklyn-grown indigo leaves by natural dye artist Liz Spencer, The Dogwood Dyer.*

*Spencer's sample book contains swatches of fabric and yarn that have been naturally dyed with indigo.*

*Spencer demonstrates how to extract indigo pigment from farm-grown indigo leaves at Maple Shade Farm in New York.*

*Spencer dyes silk in a fresh leaf indigo vat made with organic indigo leaves, lime, and fruit sugar.*

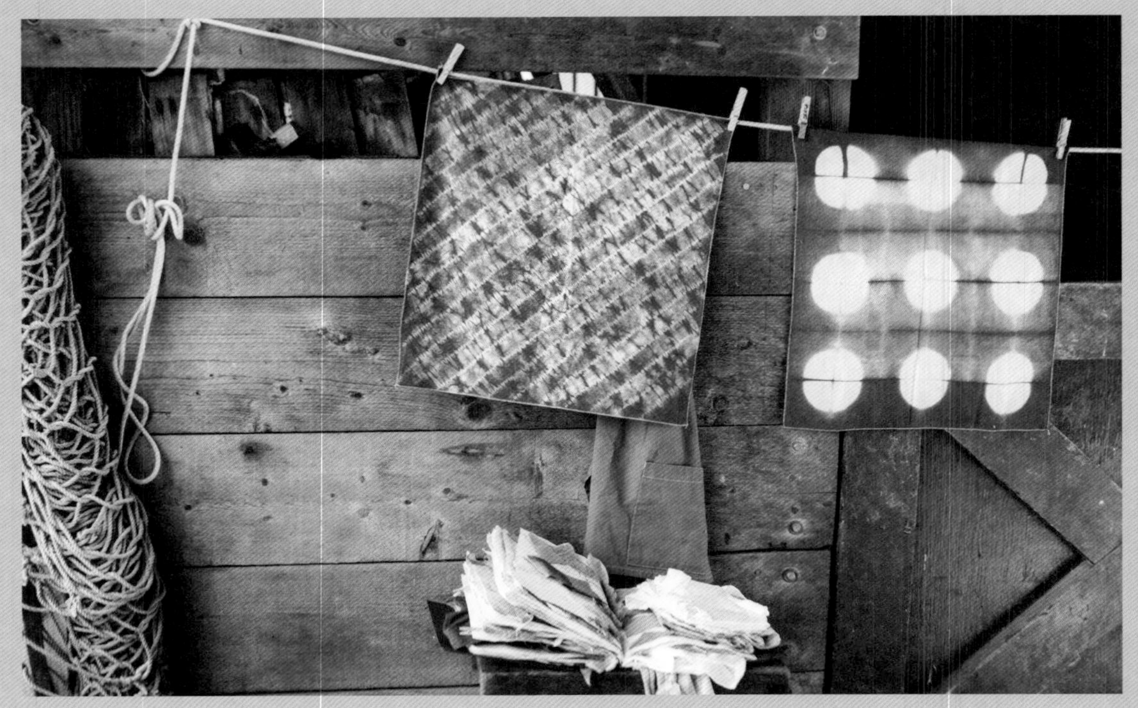

*Textiles dyed using* itajime *and stitched* shibori *techniques, both Japanese manual resist dyeing methods.*

*Swatches of fabric and yarn naturally dyed with indigo in Spencer's sample book.*

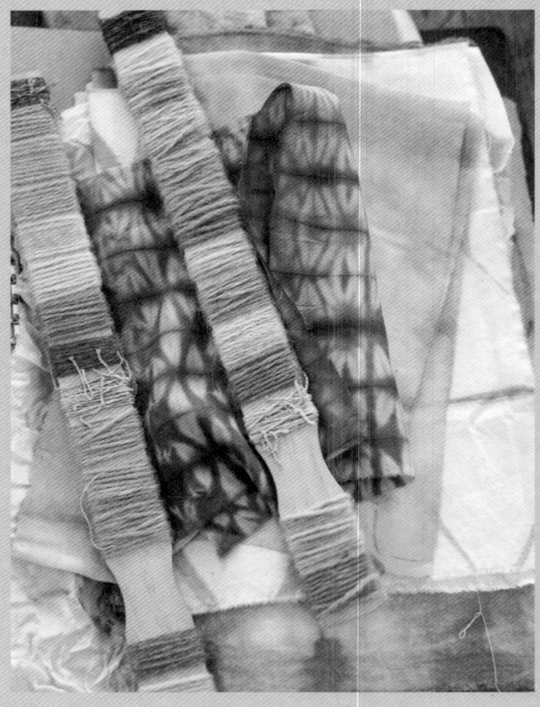

*Yarns and fabrics dyed with locally grown and foraged dyestuffs from Spencer's dye journal.*

*Spencer pre-wets and hangs organic cotton and silk scarves for indigo dye workshop participants at Brooklyn Grange Rooftop Farm (see p. 226).*

*Jars of natural dyestuffs during a bundle dye workshop with Spencer at Brooklyn Grange Rooftop Farm.*

*Onion skins saved from the kitchen of Brooklyn restaurant Lighthouse are used as a dyestuff for workshops.*

*Rolling up a bundle of natural dyestuffs on silk. Bundle dyeing produces a unique pattern every time.*

*Binding the dyestuff and silk in preparation for the steaming process. Bundle dyeing is a water-saving method.*

*Brightly colored silk naturally dyed using the bundle dye process at a Brooklyn Grange Rooftop Farm workshop with The Dogwood Dyer, Liz Spencer.*

fabric, and the clear part of the fabric is then folded over before it is rolled up tightly. Twine is needed to wrap it up and bind it.

The wetted bundle is placed on a strainer inside a pot with a little warm water in the bottom. Covered with a lid, it is then steamed for at least 30 minutes.

Indigo dye, which Spencer refers to as the "king of natural dyes" because it can make purples and greens apart from blues, requires a different process. She uses the Michel Garcia 1–2–3 indigo vat, the formula for which calls for one part indigo, two of lime (calcium hydroxide), and three parts fructose. This process works very well with cotton and other plant fabrics.

Create an average batch by adding 100 grams (3.5 ounces) of natural indigo powder, an extract of the indigo plant, to a stainless-steel stock pot full of water and mixing it up well. Then add the lime, mixing it up again. Finally, add fructose, the reducing agent and catalyzer that allows the indigo to dissolve and bond to the fabric. Wait at least 20 minutes, or overnight, and then the solution is ready for dyeing.

Using rubber gloves, soak the fabric in the indigo dye solution for one minute, then pull it out and wait for five minutes for the dye to fully oxidize. By dipping and pulling it out multiple times, you add layers of indigo to the fabric.

# Purple Saffron Fields Flourish High above the City

*Bringing color to rooftops across the French capital, BienÉlevées grows environmentally friendly Parisian saffron*

An urban farming start-up founded by Amela du Bessey and her sisters, BienÉlevées focuses on growing high-quality, environmentally friendly saffron on rooftops across Paris while offering workshops involving this intriguing spice that counts Alexander the Great among its devotees.

Often referred to as "red gold," saffron is the most expensive spice in the world. The high cost is a result of the labor-intensive process of extracting the stigmas from the flowers, and the number of flowers necessary to produce small quantities: 150 flowers will produce just one gram of dried saffron. Iran, Spain, and Afghanistan are currently the top saffron-producing countries in the world.

"This project started by coincidence actually," says Amela du Bessey, who founded BienÉlevées together with her sisters Louise, Philippine, and Bérengère. "I tried to plant it at my parents' house in the countryside in normal conditions, but I also put a few bulbs on my balcony in Paris because I wanted to check when it was going to bloom. I saw, year after year, that the plants looked more comfortable on the balcony in the city than

at my parents' home in the countryside. I was amazed by that, and I started to think how saffron would fit in urban agricultural conditions. I studied it a bit more and that's how the project started."

In 2018, the project was backed by the second edition of Parisculteurs, a public campaign that aims to cover 100 hectares (247 acres) of rooftops and walls in Paris with greenery. "We built the case, we won it, and that was very reassuring for us because in the jury there were many experts who said that what we were planning to do should work," explains du Bessey.

That same year, the sisters planted two fields: their Parisculteurs-backed plantation, on the roof of a supermarket located in the city's 13th arrondissement, and another on the rooftop of the Jean Nouvel-designed Arab World Institute, beside the Seine river, in the 5th arrondissement. In 2019, they added three more sites to their project and currently they cultivate saffron on various rooftops, with a total growing surface spanning 1,700 square meters (18,300 square feet).

*"If we want to be able to make it a profitable activity, which is not the case yet, we need to have more volume, more rooftops, and more saffron."*

According to du Bessey, saffron likes very light soil—the key is that water can flow freely. "The soil that is good for the buildings, because it's very light, is the same soil that is good for the saffron, because it doesn't retain water," she says. Last year, they planted 93,000 bulbs and they hope to harvest over 1 kilogram (2.2 pounds) of saffron, a significant increase from the previous harvest, when they collected 700 grams (25 ounces). "The magic of the plant is that the bulb you plant one year transforms into three bulbs the next year, and then six," explains du Bessey. "It doesn't reproduce with pollen; it multiplies itself in the ground."

With three harvests behind her, du Bessey is now used to hectic early autumns, the busiest time of their farming year. Come October, they visit the saffron rooftops daily to check on the vibrant purple flowers, which have to be picked by hand on the morning they come into blossom, before

*1] The four sisters and co-founders of BienÉlevées (from left): Bérengère, Philippine, Amela, and Louise on the Lycée Hôtelier Guillaume Tirel's rooftop (p. 97)*
*2] Diana plants crocus bulbs (top)*
*3] Diana and Yan plant three-year-old* Crocus sativus *bulbs, which will bloom in October (p. 99 top)*
*4] BienÉlevées gardener Denis (far right) during a saffron workshop in Montrouge (p. 99 bottom)*

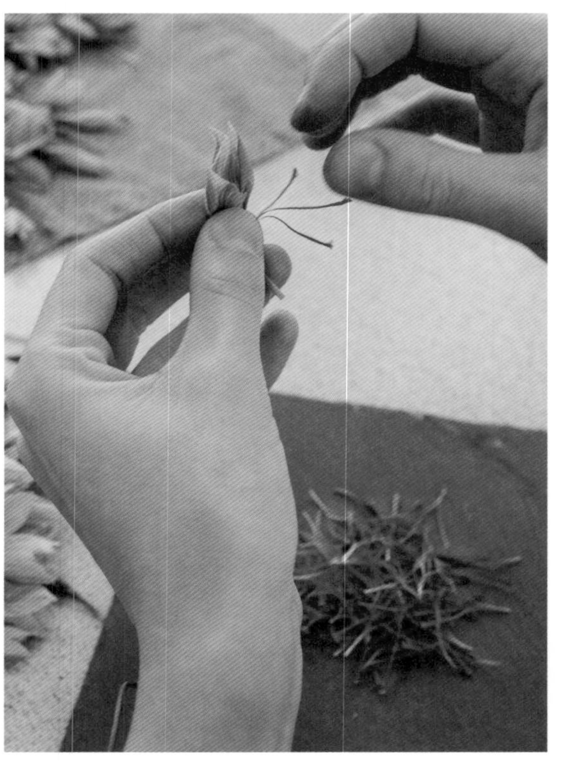

they can be damaged by the elements. Baskets full of saffron flowers are then transferred to a safe place to do the *émondage*—separating the pistil from the flower. "This is very delicate," says du Bessey. "It has to be done with strict sanitary conditions and also technical knowledge, to have the right lengths that will keep all the power of saffron but not the part that has no intensity. This has to be done on the day that the flower blooms, or it is too late."

The taste and aroma of saffron is only contained in the red-orange stigma and drying them is a key part of the process for the latter to become saffron threads. "We choose to have a very short drying period so that the flavor is strong but not very spicy. Our drying machine works with solar energy. Then the saffron has to be kept in boxes in a dark place so that the flavor won't go away, and it has to wait at least three weeks until it starts to smell like saffron. The longer we wait, the better it tastes."

They often receive requests from city dwellers looking to help them, at no charge. "We don't want to use free labor; we think it's not a healthy way of running a company. So when harvest time comes, we hire some people just for those few weeks." Like many other urban farming companies, having various streams of revenue is key to making BienÉlevées financially viable. Half their revenue comes from workshops, where the sisters share both technical knowledge and tales of saffron through the ages. (One involves Alexander the Great, who is said to have taken saffron baths to heal his battle wounds.)

Added-value products like saffron biscuits or saffron petal-dyed scarves are likewise part of their repertoire. "There are two reasons for that: one is that in France—I think it's different in Spain and other countries—people aren't used to cooking with saffron. A lot of people tell us, 'I don't know how to cook with it; I don't know what quantity to use.' So that's why we wanted to have other products like biscuits, because it's easier, it's a way of getting them used to the spice." The otherwise doomed-to-be-discarded saffron petals are used to naturally dye silk scarves, available on BienÉlevées's website together with other objects, including beautiful notebooks featuring saffron recipes.

Legal requirements are among the main challenges for BienÉlevées to become profitable. "If we want to be able to make it a profitable activity, which is not the case yet, we need to have more volume, more rooftops, and more saffron," explains du Bessey. "We have loads of building owners who are interested in the way we work and want to have saffron on their rooftops. But when we start with the legal people to formalize the contracts to use the rooftops, then it gets complicated, very long and very difficult." Despite the challenges, with plans to expand to Lyon, one of France's most exciting food destinations, and interest from other cities on the rise, the future of BienÉlevées looks bright. "Our mission," says du Bessey, "is to make cities greener and more delicious, and to bond people together."

*5] A bee investigates harvested crocus flowers. Bees pollinate the flowers and produce saffron honey (p. 100 top)*

*6] Filaments are always harvested by hand from the* Crocus sativus *plant (p. 100 bottom left)*

*7] BienÉlevées's saffron harvest is dried as soon as possible to ensure optimum quality (p. 100 bottom right)*

# The Food Ecology of Urban Beekeeping

*How beekeeping can connect urbanites to the natural world present in the city*

BY URBAN BEEKEEPER NIC DOWSE

When beekeepers gently smoke a beehive, crack open the lid, place their hands within the honeybee (*Apis mellifera*) superorganism, and taste the honey in the comb, we are joining a food web that connects us to many other creatures in the city in a meaningful way. We become connected to the marbled geckos (*Christinus marmoratus*) that often live under the pot plants we keep on the tops of hives, enjoying the warmth of the bee cluster below throughout the cooler months and snacking on bees that wander their way. We become connected to the red wattlebird (*Anthochaera carunculata*) that enjoys the same nectar in our flowering native trees as honeybees do, as well as the high-protein meal of the bees themselves—especially in spring when there are hungry mouths to feed in the nest.

Many ethical consumers in the twenty-first century have a growing appreciation of beekeeping as a farming practice that connects urban communities to the local production of food. We are becoming much more aware of the essential role insects play in pollinating the fruits, vegetables, nuts, and herbs in our backyards, decks, and community gardens. Pollination increases the yields of many favorite foods, and beekeeping assists with that process. Bringing food production back into cities reduces food miles and makes visible the skills of small-scale farming to city populations more used to buying carrots than growing them.

However, many consumers may not be aware of beekeeping as a farming practice that also connects us to the natural world of non-human species that, like humans, relies in some way on the beehive for nourishment or shelter. We often think of "nature" as being somewhere remote, far removed from the city and its dense human populations. The observant beekeeper, who watches the natural world surrounding and interacting with a beehive, realizes that the city is full of the natural world too.

We begin to appreciate that there are many species of pollinating insects and native bees that also need a habitat to breed in and food to thrive on. Holistic beekeeping recognizes this, and when gardeners ask us beekeepers what to plant for bees, we suggest flowers and crops that benefit indigenous species too. In Melbourne we know that our native blue-banded bee (*Amegilla cingulata*) loves blue flowers such as salvias and can pollinate food crops that honeybees cannot, such as tomatoes and eggplants. We also know that many native bees are solitary creatures that burrow into the ground to make their nests, so we ask gardeners to leave some sections of earth exposed and not covered in mulch that, while it may retain nutrients and water, deters many native bees from establishing nesting sites.

The holistic beekeeper is constantly observing the urban spaces within which they work, and takes mental notes of the quiet, yet

very special, interactions with birds, insects, and other non-human species. We strive to make room for the many, often overlooked, small creatures in our urban landscape.

Beekeeping, if practiced in this reflective manner, is an expansive farming tradition that acknowledges connections between regenerative farming and urban ecologies, and includes both exotic and indigenous species. When we taste the landscape in the honey gathered from the millions of blossoms in our city streets and gardens, we feel the wonder of being a little part of something much bigger than ourselves.

At Honey Fingers, we acknowledge the traditional owners of the lands upon which we beekeep today, including the Wurundjeri and the Boon wurrung peoples. We pay our respects to their Elders past, present, and emerging.

*An architect by training, Nic Dowse is a Melbourne-based urban beekeeper and founder of Honey Fingers, a creative collective that experiments with and researches bee culture.*

# The Art of Beekeeping in an Urban Environment

*Wilk Apiary manages urban and rural beehives across New York State to produce natural, local honey*

With bee populations in sharp decline across the world, urban beekeeping has flourished in the last decade. Beekeeping has become hugely popular in metropoles like New York City, where registered beehives have quadrupled since beekeeping was legalized in 2010.

Tom Wilk, of Wilk Apiary Inc., started his first hive in a community garden in Long Island City, Queens, in 2012 after taking an introductory course in beekeeping at the New York City Beekeepers Association. It was the book *The Backyard Beekeeper* by Kim Flottum—a read suggested by a Putnam County farmer—that sparked his interest to learn more.

Nowadays he is one of the most respected beekeepers in the city. He has been running the NYC Honey Festival since 2017, founded the Queens Beekeepers Guild in 2018, and earned his Master Beekeeper Certification from Cornell University a year later. Currently, this one-man-band tends nine apiaries with 53 hives in different locations across New York State—13 of them in the city, in venues ranging from a school to a brewery and a private backyard. "One major issue I see is that there are too many hives too close together for the urban environment to support," says Wilk. "Another problem is most people who start to keep bees don't really understand what they need to do. If one beekeeper within five kilometers (three miles) of your hives gets diseases in their hives, they can spread them to your apiary, even though you manage them to be disease-free."

Honeybees play a crucial role in natural ecosystems, and it is estimated that one-third of the U.S. food supply relies on them for pollination. Unfortunately, their populations are vanishing at alarming rates. Colony collapse disorder (CCD), a mysterious phenomenon in which worker bees abandon their hives, is often cited as a cause, as are pesticides, pathogens, and monoculture, which limits the forage supplies for bees.

"In the city, you worry less about pesticide use and you don't have to worry about monoculture—there is a wide variety of flowers the honeybees can go to for nutrition," says Wilk. "The main source is when the trees bloom, as one nice-sized tree can have close to half-a-million flowers." In his urban apiaries, he hopes for 13.5 kilograms (30 pounds) of honey per hive, but his suburban hives can sometimes produce up to 45 kilograms (100 pounds). However, some don't produce enough extra honey to harvest.

Wilk's hive work, scattered across the city, is distinctly seasonal. While in the winter he

passes by once in a while to ensure the entrances to the hives aren't blocked by snow, spring and summer require more dedication. "In the spring, we start by installing packages of honeybees into hives or splitting the hives that survived over the winter," he says. "Splitting makes one hive become two hives, and when done at the right time, it will keep the survivor hive from swarming in May. I visit each hive weekly and make sure they have food until the natural bloom starts and they can get their nutrition naturally."

Swarming is the bees' natural method of propagation. It happens when the colony becomes crowded and the queen bee leaves the hive with a group of worker bees to form a new colony elsewhere.

In the summer, Wilk visits less often, trying not to disturb the bees at peak season. As the number of bees in a hive increases—they can be home to up to 50,000 bees—he adds honey supers so the bees have plenty of room for the extra nectar they are bringing in, and to prevent swarming, which could mean losing half the bees overnight.

*"In the city, you worry less about pesticide use and you don't have to worry about monoculture—there is a wide variety of flowers the honeybees can go to for nutrition."*

*1] Tom Wilk inspects two of his hives at the Long Island City Roots Community Garden in Queens (p. 105)*
*2] Wilk prepares his smoker for the hive inspection at his Hellgate Farm Apiary, Queens (top)*
*3] Hive inspection with beekeeper Geraldine Simonis at Wilk's Ridgewood Rooftop Apiary, Queens (p. 107 top)*
*4] Frame with honey and bees at Wilk's Finback Brewery Apiary, Queens (p. 107 bottom left)*
*5] Frames of honey sit in the honey extractor at Wilk Apiary, Queens (p. 107 bottom right)*

The first harvest usually happens in July. "I call this the spring honey," Wilk says. "It is lighter and has a predominance of the largest bloom so far, which is the linden trees." Treatment against *Varroa destructor*, an external parasitic mite that spreads viruses to the hive, is usually conducted after the harvest. The second nectar flow begins in mid-August through September. Sometimes, Wilk gets an early autumn harvest too. "When I harvest honey, I use an extractor that uses centrifugal force to spin the honey out of the comb, so the bees can reuse the comb. I then gently strain the honey and bottle it for sale." He bottles and markets the single-harvest New York City honey based on the hives' location, adding a zip code on the jar—his 225-gram (8-ounce) jars sell for $25.

"Once the nectar flow stops, I check to see that there is sufficient honey in the hives for them to get through the winter," explains Wilk. "If not, I feed them sugar syrup with supplements. Before it gets cold I winterize

FINISHED
COMPOST

my hives by wrapping them and adding a moisture barrier inside the top cove, because condensation running over the bees during the cold weather would kill them."

Wilk never intended to go into full-time beekeeping, but he did investigate the possibility when he lost his day job due to the Covid-19 pandemic. He concluded it would be impossible to support his family with beekeeping alone. "From the few beekeepers that I know doing so," he says, "I see that they have to make concessions in order to make more money—teaching people who really shouldn't be keeping bees, or squeezing as much honey as they can out of their hives and not leaving enough honey for the bees to use. They can end up taking shortcuts to save money on things like mite treatments and winterizing their hives."

For Wilk, treating bees as living beings and caring for their safety and well-being are essential for beekeeping. Since he became a beekeeper, he has made it his mission to raise honeybees in New York City in order to help the environment and share his knowledge. "Educating the public is one of my favorite things to do," he says. "I want to take away the prejudice and fear of honeybees while showing people how they help the environment. Pollinators in general are important to the environment because they pollinate flowers, fruits, vegetables, and trees, enabling the plants they pollinate to pass on their genetics and help clean the air. Honeybees are important because they are the only pollinators that can be raised relatively easily and benefit us all with honey, pollen, propolis, and beeswax."

*6] Wilk manages his beehives at the Long Island City Roots Community Garden (p. 108)*
*7] Pointing out the queen during a hive inspection (p. 110 top)*
*8] Simonis uses a smoker to calm the bees before inspecting Wilk's Ridgewood hives (p. 110 bottom)*
*9] Wilk Apiary's Queens Blossoms honey is harvested from multiple apiaries and blended together for a true taste of Queens (top)*

# How to Start Beekeeping at Home

*An ancient art that has seen a resurgence, many urbanites are setting their own beehives and harvesting their own honey in the city*

Before setting any beehives, it's important to make sure that there is enough forage around for the bees to feed on, and that local regulations allow the activity.

Managing bees is a ritual that demands care and knowledge, and that's why many beekeepers decide to join courses or read specialized literature on the subject before they begin their beekeeping journeys.

The first piece of equipment you need is a beehive—the most common ones are usually the Langstroth type. Ideally, the hive should be placed in a sunny spot with minimal wind and easy access to a water source for the bees.

According to Peter von Ziegesar and Hali Lee, a starter kit should include protective gear—especially a veil, hat, and gloves—and a smoker,

*Beekeepers Hali Lee and Peter von Ziegesar light the bee smoker used for hive inspection.*

*Von Ziegesar holds a honey-filled frame containing worker bees. As their name suggests, they are responsible for all the work in the hive.*

*Worker bees have a life span of only four to six weeks.*

*Capped brood cells, the last stage before baby worker bees are hatched.*

*Von Ziegesar and Lee pull a frame out of the hive for inspection, a key step in detecting any potential diseases.*

*A smoker used to calm the bees sits on one of the sealed rooftop hives.*

*View of downtown Brooklyn from three of von Ziegesar and Lee's rooftop hives.*

an essential tool when inspecting the beehives since the smoke makes the bees feel calmer.

Catching a bee swarm might be too ambitious for beginners, so usually the easiest method is to buy bees for the new colony from a respectable source. Likely, you will receive the queen bee in a separate container.

Once you have placed all the bees inside the beehive, let them alone for at least a week to adapt to their new environs, ensuring the feeder is full during that time.

The warmer months are high season for bees and that's when honey is harvested. A crowbar-style hive tool might be handy to open the hive and pull the frames out.

Once that is done, uncap the wax-sealed honeycomb on both sides of the frame with an uncapping fork and place the frame in a honey extractor. Straining is the last step before bottling the honey.

Von Ziegesar and Lee, who have been keeping beehives on their Brooklyn rooftop for over five years, recommend renting or borrowing the extractor—the most expensive piece of beekeeping equipment—especially at the beginning.

Keeping an eye on pests is crucial—Verroa mites are the most serious one—since early detection can prevent further damage.

During the colder months, it's necessary to make sure that the beehives are winterized and the bees have enough food to survive.

# Design and Innovation

*New technologies surface every year, creating endless possibilities for innovative farmers willing to implement them*

The idea of software engineers growing vegetables without sunlight in immaculately clean warehouses, or strawberries protruding from towering structures without soil, might have sounded like science fiction some decades ago, but today pioneering farms—spaces that could often pass for art installations—are taking center stage.

That's the case with Brooklyn-based Smallhold, an innovative farming operation growing organic mushrooms in controlled-environment mini-farms designed in-house. They can also adapt their high-tech farms to any space, installing them in establishments ranging from restaurants to grocery stores and hotels, honoring the real significance of freshness and revolutionizing the food supply chain in the process.

The company, founded in Brooklyn by best friends Andrew Carter and Adam DeMartino, grows an array of sculptural, lesser-known mushroom varieties, including pink oyster, maitake, lion's mane, and royal trumpet. "Smallhold uses proprietary technology and a sophisticated supply chain to spin-up small-to-mid-scale local farms in regions around the world," explains Carter, the company's CEO. "All facilities and mini-farms are connected through our hardware and software system, which allows us to share resources, raw materials, data, and expertise, enabling local, organic production at a competitive price point."

With their ability to design, build, and install the distributed bespoke mini-farms they grow their rolling supply of fresh mushrooms in, they have created a novel fungi kingdom that functions seamlessly. Smallhold's innovative approach brings two previously alien concepts—farm to table and tech start-up—together, in order to make a bounty of fresh mushrooms available in produce aisles, restaurants, and even private homes through their growing kits.

"Most mushrooms are not easy to grow on a commercial scale; they're considerably harder than plants," says Carter. "To give us the consistency our customers desire without using any pesticides requires extreme control over all environments and our supply chain." They have achieved the accurate control system required to grow such delicate crops successfully in different locations by developing FarmLinc, their own hardware and software system, which tracks all climate parameters as precisely as a surgeon. The system allows the farming units to be largely automated and connected to a server, which lets them access the units from any laptop, anywhere.

*"Commercial indoor farming is tailored to hobby or massive scale. Our target footprint and method doesn't exist."*
*—Andrew Carter, CEO of Smallhold*

In a similar vein, Higher Ground Farm demonstrates how interconnected technology can be applied to create more sustainable and efficient farming conditions, the impact of which goes far beyond the production of food. Based in Boston, this venture is focused on implementing rooftop farms atop healthcare institutions, not only providing fresh, organic food, but also tangible educational and team-building opportunities within the community. One of these projects finds its home at the Boston Medical Center. Run by Lindsay Allen, Higher Ground Farm's farming operations director and BMC's farm manager, the rooftop farm was designed using 2,300 milk crates and has a water-efficient, high-tech irrigation method that Allen controls from her phone. A significant amount of the produce grown on the rooftop farm goes directly to the preventive food pantry; the rest goes to the hospital kitchen, a teaching kitchen, and to a farmers' market. The benefits of this innovative system can be felt in other fields too. Only three stories up, the farm is visible from one of the hospital's main clinics. "Studies show that having visual access to green space can reduce employee stress and recovery time for patients, so we wanted to make sure the farm was visible to as many people as possible—that's why we placed it where we did," says Allen. The rooftop greenery also helps absorb stormwater and reduces the heating and cooling needs of the offices below.

Innovation in farming is often interconnected with design. Another way of exploring modern approaches to urban farming is by growing vegetables in a soilless environment, using inventive methods that rely on a nutrient-water solution for plants to feed on. Agripolis, a French urban farming start-up, designs and builds their own farms in the open air, on rooftops in urban areas. Their mostly vertical aeroponic and hydroponic farms don't need weighty soil to thrive—a creative solution to the limited load-bearing capacity that buildings present.

The Porte de Versailles farm Nature Urbaine, conceived and operated by Agripolis, is located on top of a major exhibition complex in southwestern Paris. Considered the largest rooftop farm in Europe and one of the largest in the world, more than 20 varieties of fruit and vegetables are grown using ultra-modern, lightweight aeroponic

*New technologies are allowing urban farming ventures to distribute their growing spaces across different locations, both indoors and outdoors. Nature Urbaine's aeroponic vertical farm at Paris Expo Porte de Versailles is operated by Agripolis (p. 117). Smallhold grows royal trumpet mushrooms at their Brooklyn farm (p. 116) and have installed an automated, custom minifarm over the bar of New York hotel The Standard, East Village (p. 119 top left); lead farmer Louis transports the Blue Oyster mushrooms to the kitchen of the hotel's Café Standard (p. 119 top right). Strawberries are picked at the Agripolis rooftop farm above the Cour des Lions swimming pool in Paris (p. 119 bottom).*

and hydroponic installations. Their closed-water system feeds a nutrient-rich, misty water solution to plants that grow in cleverly devised hollow columns suspended from a bamboo structure, a design that maximizes the space available. They use the same aeroponic and hydroponic system at their considerably smaller Piscine Cour des Lions farm, set on top of a swimming pool in the 11th arrondissement.

*"Our system tracks all climate parameters and collects millions of data points per day across the network." —Andrew Carter, CEO of Smallhold*

"Design and innovation play a central role in the farm model we sell," says Pascal Hardy, founder of Agripolis and a specialist in urban agriculture. "Particularly with issues like closed-circuit cultivation systems, with monitoring and alarm systems that provide full control of cultivation parameters and allow the farm to be controlled from a distance, and also with the conception of our cultivation systems, like towers or bamboo structures."

Urban farming projects can infuse new life into offices and buildings, giving added value to work spaces. That was the case with the now defunct Pasona building's urban farm in downtown Tokyo, an oasis with a green facade and multiple farming facilities integrated across different floors within the building. Conceived by Kono Designs, the innovative farm included 200 species of fruit, vegetables, and rice, and allowed employees to harvest their own crops at the office.

The design and innovation landscapes in urban agriculture are diverse and dynamic, covering all aspects of the food production process, from distribution to maintenance work, and of course the actual growing of food. New technologies provide an ever-evolving palette for urban farming start-ups to experiment and expand alternative growing methods. Diversifying might prove particularly critical in the upcoming decades, when innovative solutions for growing more food with less natural resources are likely to become more essential than ever.

*Higher Ground Farm at the Boston Medical Center (p. 121 top) uses high-tech solutions such as an irrigation system with root emitters (top) and a Weathermatic SmartLine controller, set here by farm manager Lindsay Allen (p. 121 bottom left)—she can also control the system remotely. Farm intern Christine washes collard greens (p. 121 bottom right) at Higher Ground Farm, where food is grown for patients, visitors, staff, and their on-site Preventive Food Pantry.*

# Growing and Selling Living Produce across the World

*Born in Berlin, Infarm's successful indoor farming model is expanding worldwide*

Infarm is a highly efficient vertical indoor farming company that deploys modular farming units inside supermarket aisles and restaurants to bring their fresh living produce closer to the final consumer. Their leafy greens and herbs are grown under artificial lighting hydroponically, meaning plants are cultivated following a soil-free method and fed on a blend of water and nutrient-rich solution. Environmental factors are not an issue for Infarm: using controlled-environment agriculture technology, they can create ideal conditions in which temperature, lighting, humidity, and even nutrients are monitored and adjusted according to the plants' needs.

Founded in Berlin in 2013 by Osnat Michaeli and brothers Erez and Guy Galonska, the company's modular vertical farms—a distributed model that is easily scalable—can be found on the aisles of supermarkets across the world, from Marks & Spencer in the United Kingdom to Kroger in the United States, or Metro and Aldi in continental Europe. Their business model is based on offering retailers the plants themselves so that customers can buy the living produce: Infarm provide the technology and the farm; they grow the plants and harvest them.

"We started with the simple idea that we wanted to make fresh, pure, tasty, and nutritious produce available and affordable for everyone," says co-founder Osnat Michaeli, Infarm's chief brand strategy officer. "We decided we needed to redefine the entire supply chain from start to finish to make this a reality for those living in urban environments. We wanted to bring nature, and the benefits of fresh produce, fully into city life."

The realization that as a society we are removed from the food-growing process and the harm it can cause—that we tend to know little about how food gets from the field to our plates—fueled the trio's decision to distribute farms throughout the city, rather than building them outside urban centers. "In a phase of trial-and-error experimentation in 2013, we converted a 1955 Airstream trailer into our first vertical farm in one of Berlin's most well-known urban farm spaces, Prinzessinnengarten," says Michaeli. "That trailer became a hub of activity for a range of professions, from urban planners and designers to food activists and biodynamic farmers, all of whom we invited to explore with us the diverse challenges behind making urban farming a reality. In doing so, Infarm was born."

infarm

infarm

HERBES AROMATIQUES FRAÎCHES
SAVEUR PRÉSERVÉE
infarm
METRO

Their growing system includes patented technologies and cutting-edge research and development, from plant science to data science and engineering. Over 50,000 data points—covering factors such as humidity or temperature—are collected and transmitted to the company's cloud, allowing Infarm to control the growing conditions of their plants remotely, recalibrating as needed across their network to optimize their growing methods.

As with many other businesses across most industries, they had to adapt quickly to the unpredictable circumstances triggered by the global health emergency, which brought new challenges daily. But their farming model proved resilient, allowing them to respond quickly to the needs of their clients, who were themselves establishing new procedures for producer and customer interaction in their retail spaces. "Across our global network, each of our farms is connected to the cloud, allowing us to monitor plant performance and growing environment, and make adjustments remotely," says Michaeli. "We developed a system to guide local teams through farm installation remotely when our installation teams were unable to travel, further minimizing any potential impact from the changing pandemic environment to our operations." In September 2020, the company announced that it had raised $170 million in investment in the first close of a Series C funding round.

According to Global Market Insights data, the vertical farming market is expected to multiply by almost seven by 2026, reaching a projected value of over $22 billion. Infarm expects to reach more than 46 hectares (114 acres) of urban land by 2025, and their aim is to become the largest distributed urban vertical farming network in the world.

"Our goal has always been to encourage cities to become self-sufficient in their food production, and this will not change," Michaeli says. "We have secured an incredible amount of funding, which we will use to deepen the regional and local penetration of Infarm's global farming network and complete the development of Infarm's new generation of vertical cloud-connected farms. These new farms will save labor, land, water, energy, and food miles, contributing to a more sustainable food system. We want to continue practicing a form of agriculture that is resilient, sustainable, and beneficial to our planet, making fresh, pure, tasty, and nutritious produce available to everyone."

*1] Infarm employee Rodrigo at harvest time at warehouse METRO Nanterre (p. 123)*

*2] Rodrigo photographs harvested plants to help track plant growth and health, and improve production (p. 124 top)*

*3] Infarm's vertical vegetable garden produces nearly four tons of aromatic herbs per year, which are sold on-site at the METRO store in Nanterre, France (p. 124 bottom)*

*4] Du, an Infarm employee, wraps parsley to be sold. The roots help keep the herbs fresh for longer (top)*

# A Mushroom Farm in Paris

*A formerly vacant car park in Paris's 18th arrondissement is home to subterranean mushroom farm La Caverne*

Located in a disused car park in northern Paris, La Caverne is an underground farm growing organic mushrooms and Belgian endives. Founded by urban farmers Jean-Nöel Gertz, a thermal engineer by training, and Théo Champagnat, an agricultural engineer and nomadic chef, this is the first certified organic farm in the city, with a mission to produce local, organic, carbon-neutral vegetables.

The work to transform the abandoned multistory car park into an underground farm began in late summer 2017, after the pair won a call by the city council for proposals to make Paris greener (the first edition of Parisculteurs). This initiative, currently in its third edition, aims to convert buildings across the city into hosts for green ventures.

Since the official launch of the farm in spring 2018, the amount of space devoted to growing vegetables in the reconverted car park has more than doubled. At first, Gertz and Champagnat also cultivated microgreens following hydroponic methods, which use water and a nutrient solution instead of soil, but since organic certifications aren't currently extended to non-soil growing methods in the European Union, the duo decided to focus exclusively on mushrooms and endives, both suitable for cultivation in the dark. At present, they are cultivated in more than half of the 17,000 square meters (183,000 square feet) available at the farm, and the plan is to use all the existing surface area in the next two to three years.

*Every week, over 800 kilograms (1,760 pounds) of vegetables are produced at La Caverne, including shiitake, button, and oyster mushrooms.*

While the Belgian endives have a two-phase growing process (the roots grow in open fields in the countryside for six months before being harvested and transferred to the farm, where they spend their last three weeks developing in the dark), the mushrooms complete their full growing cycle on-site, in substrate blocks that sit on metal racks hanging from the ceiling, a clever design that optimizes space.

Paris's love affair with mushrooms is part of its cultural heritage. In fact, button mushrooms are known as *champignons de Paris* (mushrooms from Paris) in French.

By 1875, more than 1,000 tons of these were grown each year in the dim network of limestone quarries beneath the city. Historians often associate the demise of the Parisian mushroom farming tradition with the development of Paris's underground metro system at the end of the nineteenth century.

The company's unusual choice of venues has unearthed other ties with the past: Cycloponics, the urban agriculture start-up founded by Gertz and Champagnat and to which La Caverne belongs, was born in the city of Strasbourg, on the edge of the French-German border. The pair's first urban farming project—where they also grow organic mushrooms—Bunker Comestible (Edible bunker), is located there. As its name indicates, the farm sits in a former bunker used as a gunpowder warehouse after the city fell to the German Empire following the 1870 Franco-German War.

*"Like any new business, you have to learn everything from scratch, from the people you hire, to the distributors, to farming. Everything has to be in order and the process has to be strict."*

"I was working as a building inspector, but in 2008, with the economic crisis, I couldn't find any jobs," says Gertz. "That's when I began to think about changing my career and doing something that I love, like farming. But I didn't own any fields, so I had to look for some space, and I searched in the city center. In Strasbourg we have a lot of underground facilities because of German and French disputes over the last few centuries—the current border was re-established after Nazi Germany's defeat in the Second World War—and so a lot of bunkers were built. I knew there were spaces underground and I found one available, built by the Germans. I asked the city council if I could do something with it." The local authorities accepted his proposal and Gertz and Champagnat's first urban farming project was set up there.

Each founder brought different expertise to the project: "We had everything we needed to work underground—my building

*1] Théo Champagnat, co-founder of La Caverne, harvests oyster mushrooms (p. 127)*
*2] La Caverne grows organic mushrooms and endives in the former Raymond Queneau subterranean car park (p. 128 top)*
*3] Shiitake mushrooms emerge from growing blocks (p. 128 bottom)*
*4] Champagnat harvesting organic oyster mushrooms (top)*

skills, and Théo's agronomist skills," explains Gertz. The early days were challenging, mostly because everything was new: "Like any new business, you have to learn everything from scratch, from the people you hire, to the distributors, to farming," explains Gertz. "Everything has to be in order and the process has to be strict." But the lack of pressure from real estate companies is, according to Gertz, one of the great advantages of working underground.

*"We try to be smart, work fast, and save as much energy as we can, while we produce local, organic, carbon-neutral vegetables."*

Today, they sell their produce to super markets and small grocery stores, with the occasional customer popping in directly to the farm. With freshness in mind, goods are always delivered on the same day they are harvested, mostly by bike. "We try to be smart, work fast, and save as much energy as we can," says Gertz.

*La Caverne is bringing meaningful jobs to a disadvantaged area with one of the lowest average household incomes in the city.*

As well as providing fresh local veg, La Caverne, which is set under a social housing complex in the western side of the 18th arrondissement, is bringing new meaningful jobs to a disadvantaged area with one of the lowest average household incomes in the city.

Currently, they still have plenty of unused space, and so they partner with other food start-ups—from microgreens-growing small businesses to a start-up that packs fruit and veg boxes—to optimize the space. And with a new location in Bordeaux and other French cities including Lyon following suit, Cycloponics are expanding their underground urban farming model across the country.

*5] Boubakar manipulates organic endive shoots before moving them to the dark forcing room for three weeks (p. 130 top)*
*6] Renzel picks organically grown endives to cut the roots (p. 130 bottom)*
*7] Champagnat weighs oyster mushrooms to complete an oder (top)*

# Accessible and Open, the Vertical Farms of the Future

*Born in the heart of Manhattan, Farm.One's vertical farming model is bringing unusual ingredients to chefs and customers across the city*

Located in New York City, Farm.One is a vertical farm company growing specialty produce hydroponically. Aiming to grow unusual crops for chefs across the city, they established their first experimental vertical farm in 2016 at the Institute of Culinary Education in downtown Manhattan.

Their flagship farm is located in the city's Tribeca neighborhood, underneath Atera, a two-star Michelin restaurant. In this underground space, they grow hundreds of different varieties of microgreens, herbs, and edible flowers vertically. "Our mission is to surprise and delight with fresh, local, specialty ingredients grown at innovative farms around the world," says Rob Laing, Farm.One's founder and CEO.

With a graphic design and entrepreneurial background, Laing didn't have any previous experience in the food industry before launching Farm.One. "Personally, I discovered that eating a lot more whole and plant-based food was good for me. I got excited about cooking plant-based food and I was lucky enough to take some culinary classes and discover specialty ingredients," he says. He soon realized that the rare fresh ingredients he coveted weren't easy to find in cities with a climate like New York, and when they were available, they were often sourced from faraway places. "I thought I could use this new technology of vertical farming to grow rare ingredients in the city, which would obviously reduce emissions from transport and provide fresher produce, but would also be really interesting and fun."

Farm.One favors an accessible and transparent farming model, which is visible at the vertical farms they have scattered across New York City, from the subterranean flagship in Tribeca to installations in supermarkets including Whole Foods or spaces such as Eataly NYC Flatiron.

Plants grow indoors under artificial lighting, combining LED technology, which replaces sunlight, with hydroponics, a soil-free growing method that uses a mixture of water and mineral nutrient solution to feed the plants. The farms use two different hydroponic methods: deep water culture and flood and drain. While the efficiency of LED lights has notably improved in the last decade, Laing is conscious that power usage is the main carbon footprint of indoor vertical farms. "The air-conditioning system needs power, the construction uses materials... We need to think of the life cycle of things, and we need to use renewable sources of

farm.one

power to make sure we can create carbon-neutral farms from start to finish," he says.

Before Covid-19 struck last year, Farm.One's production model was grow to order, with a focus on bespoke chef sales. "The reason we started to do that is we saw a lot of new vertical farms getting into difficulty because they would grow a lot of stuff and then they wouldn't have a customer for it," explains Laing. "It seemed very unpredictable, and we wanted to do something where we had more certainty—to know that if we planted a seed, it was going to be sold, and this was a great way to start."

Farm.One's vast seed catalogue is informed by the city's multiculturalism: chefs offering a diverse range of cuisines and styles approach them with different requests for what to grow, which in return helps them discover new produce. Laing particularly enjoys giving chefs tours of the farm. "Many of the young chefs who grow up in the city go to culinary school and never have the chance to visit a farm. So, part of our job is to show them what's possible and get them excited."

With many New York City restaurants struggling or closing down due to Covid-19, Farm.One has had to transform their production model in search of alternatives. Their new subscription scheme now delivers microgreens, herbs, and edible flowers to individual customers. "We knew that some consumers would be interested in what we were doing, but we had to completely change what we grow," says Laing. "Our subscription is going pretty well, so we are going to continue with that." Other revenue streams, including farm tours and classes, are on hold for now, but they hope to revive them in the future.

Laing considers sharing their knowledge about the diversity of food an important part of their mission. "If you think about it, people's diets in America are made up mostly of seven different plants, including corn, wheat, and soy," he says. "But there are so many different plants available, and the more we can incorporate interesting ingredients into our diet, the healthier our diet will be and the healthier the planet will be."

Mexican tarragon is among Laing's favorite crops. "It produces a little yellow flower, has dark leaves, and an anise and mint kind of taste—really striking," he says. Another uncommon herb that Farm.One grows

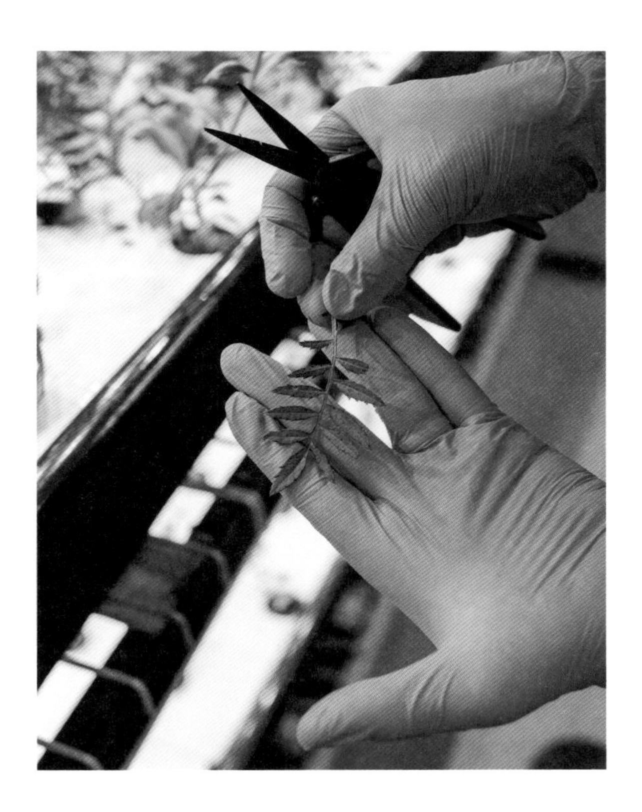

*1] Farm.One's operations director Justin Randolph holds a variety of harvested edible flowers (p. 133)*
*2] Farm.One is designed with a movable storage system that allows for more growing space (p. 134 top)*
*3] Specialty herbs, microgreens, and edible flowers are the company's primary crops (p. 134 bottom)*
*4] A freshly harvested edible marigold leaf (top)*

is papalo. "I first tasted it at the farmers' market in Santa Monica, in Los Angeles," Laing says. "It's from Central Mexico and it's got this very complex flavor—a little bit like cilantro, pepper, grapefruit—with beautiful round flat leaves."

A relevant part of Farm.One's business model is to design, build, and often maintain vertical farms, providing the technology and know-how for different companies. Their portfolio includes various projects across Manhattan and beyond, in places including Chicago and Singapore. "As long as we are building something that is accessible, visible, open, transparent, and that fits with our ethos, it can take many different forms," says Laing of their expansion. "The real difference between what we build and what some other companies build is that Farm.One farms are there to be experienced—people can visit, they can see and taste anything. That's very important to us."

Ensuring the future of vertical farming is accessible and open, integrated into the community rather than dominated by secretive corporations, is one of Farm.One's long-term goals; another is bringing down the cost of building vertical farms so that vacant urban spaces such as basements or rooftops can be used to produce food for an ever-growing urban population.

*5] Purple oxalis is an edible plant related to sorrel, with a tangy, sour taste and citrus overtones (p. 136 top)*

*6] Farm.One produces microgreens including micro arugula (pictured), micro mustards, and micro kaiware (p. 136 bottom)*

*7] Prepping orders of bachelor buttons, dianthus, and assorted marigold flowers (top)*

# GREG KIMANI

*The environmentalist tackling food insecurity by teaching urban farming skills and building edible gardens in Nairobi*

Greg Kimani is full of enthusiasm, his passion for ending hunger contagious. In his hometown of Nairobi, he is involved in various initiatives that share an ambitious goal: to give every urban resident in Africa, especially the most vulnerable, access to a healthy diet.

Food insecurity is an issue that Kimani experienced first-hand at an early age. He grew up in Baba Dogo, a neighborhood of Nairobi where poverty, unemployment, and insecurity are rampant. "There was never enough food in our family. We were many and we couldn't afford three meals a day—maybe breakfast and supper, but lunch was not guaranteed," he says. "I realized this was the main challenge. That was the norm even in our neighbors' homes—people couldn't afford three meals a day. So that's how my interest in food security started."

Kimani's grandparents had road fringe farms in Kasarani, an area located around three kilometers (two miles) from where he lived in northeast Nairobi. They planted crops such as beans and maize, and fueled an interest in urban agriculture in their grandson. "I started to think of solutions for food security because my grandparents chose to give us some food for a period to supplement our diets," he says. "That's when I realized that we can do some form of urban agriculture if we can utilize the limited spaces we have at home to grow food."

*"There was never enough food in our family. We were many and we couldn't afford three meals a day. I realized this was the main challenge."*

After gaining a degree in environmental conservation, Kimani joined the Mwengenye Lifestyle CBO, a community-based organization in Kayole, a low-income residential area. Founded in 2017, it aims to provide a wide range of opportunities for Nairobi residents who want to learn how to grow their own healthy foods in limited spaces. Kimani helped found the organization's urban agriculture resource and information center. "The aim was to provide a solution to food insecurity affecting people in urban areas, focusing on capacity building for urban residents, food security, and urban agriculture," he says.

The facility is a bustling center that includes a demo farm featuring urban farming

Motto
Educate, Enable, Empower.

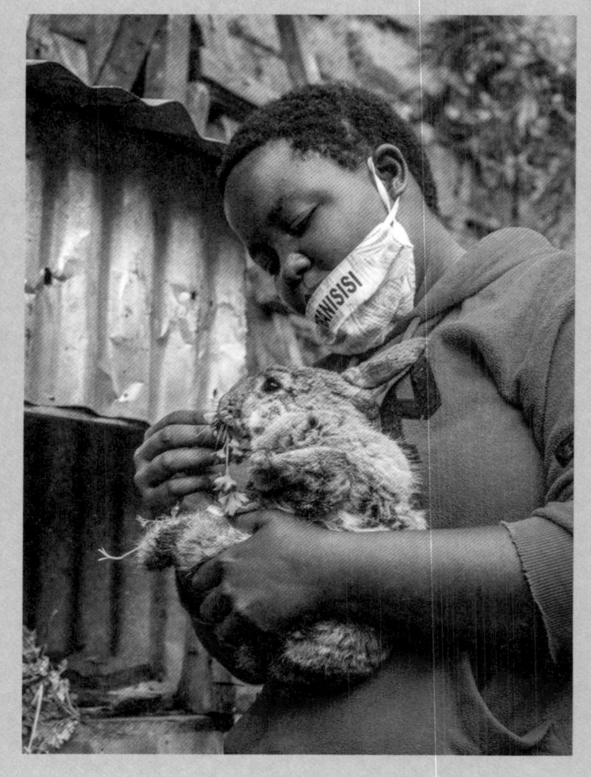

*1] Greg Kimani showcases lettuce grown in recycled plastic bottles at his Mwengenye Greens Urban Agriculture Resource and Information Center (p. 139)*
*2] Vegetables grown on a rooftop using sacks and recycled plastics as grow beds. Greg supplies households with seedlings and manure (p. 140)*
*3] The Mwengenye community garden with indigenous vegetables grown in recycled tires. Cone gardens maximize yields per unit area (p. 141 top)*
*4] Maxmillah Waithera at her kitchen garden, where she grows kale, spinach, tomatoes, onions, and herbs (p. 141 bottom)*
*5] Mwengenye Greens Urban Agriculture Resource and Information Center (p. 142 top)*
*6] Kimani and Alvin, a community volunteer, distribute grow bags to vulnerable families during the Covid-19 pandemic (p. 143 bottom left), while MaryAnn Muthoni feeds rabbits at the Mwengenye Greens center (p. 143 bottom right)*
*7] David Warue grows kale, spinach, and herbs on his rooftop with seedlings and organic pesticides from Mwengenye Greens (top)*

#TIBA
TIBANISISI

techniques such as vertical gardens. Youth organizations, women's groups, and educational institutions are its main targets. The training curriculum focuses on transforming mindsets by introducing visitors to urban agriculture, touching on topics such as horticulture, organic pesticides, waste segregation and management, hydroponic farming, and dairy goats.

*"Farming with my grandparents from a young age was an opportunity to learn indigenous knowledge from the community."*

Initially, they used the resources available within the community to set up their center, but Kimani is hoping to find investors in order to buy dairy goats and set up a hydroponic system. "I am trying to get more partners on board so that we can improve the resource center and also conduct research that can inform policy," he says.

Mwengenye Greens is the business arm of the organization, created in 2019 with funding from Nairobi-based NGO Somo Africa. Through it, the company builds edible gardens across the city, which helps tackle food insecurity while providing valuable opportunities for youth employment.

*"I am trying to get more partners on board to improve the resource center and conduct research that informs policy."*

Kimani has also built a small kitchen garden for his grandparents so they can grow vegetables nearer their home. "Farming with my grandparents from a very young age was an opportunity to learn indigenous knowledge from the community—they know a lot about plants," he says. "And now when members of the community come to the resource center, we also learn from them."

Since the Covid-19 crisis began, Kimani has noticed an increasing interest in urban farming, not only among the lower-income population but also among middle classes and high earners. His goal is to have an urban agriculture resource center in every constituency in Nairobi. "We are going to create more jobs for people in those areas where the resource centers are," he says. "Also, when we are done with Nairobi, we want to expand to other urban areas like Mombasa or Nakuru, because once the project works here, it can be replicated in other towns, not only in Kenya, but across Africa."

*8] Kimani demonstrates hydroponic farming to a group of teenagers from his community (p. 144 top)*

*9] Little Barbra admires some vegetables in the rooftop kitchen garden (p. 144 bottom left)*

*10] Demonstrating how to transplant basil at the Mwengenye Greens Center (p. 144 bottom right)*

# Empowering Women through Sustainable Urban Agriculture

*In the Indonesian town of Cirebon in Java, Konekroot is an innovative urban farm created by women, for women*

Konekroot is a small-scale aquaponic, hydroponic, and soil-based farm located in the coastal city of Cirebon, on the Indonesian island of Java. Founded by Rahma Nur Adzhani, a renewable energy enterprise and management graduate in her late twenties, Konekroot's mission is to empower women by connecting them to sustainable food production in an urban setting.

The farm, launched in fall 2019 atop a health center, spans 600 square meters (6,500 square feet) and is divided into two spaces: 400 square meters (4,300 square feet) on the fifth-floor rooftop, where a greenhouse is also located, and a 200-square-meter (2,200-square-foot) plot in the backyard.

It was a permaculture workshop that ignited Adzhani's passion for farming. "Learning about permaculture in Bali changed my philosophy of life," she says. "I wanted to grow my own food and to have the freedom to work for myself, and for nature, and that workshop made me realize that maybe—only maybe—I could do it."

During her master's studies in Newcastle, England, Adzhani had the chance to volunteer at different farms. "I had a lot of time off and I felt that learning from books wasn't real—I needed more confidence," she says. "Finding volunteering opportunities in the U. K. is so easy. Here in Indonesia it isn't, and I don't come from a family of farmers, so I did a lot of volunteering while I was there."

*"Learning about permaculture in Bali changed my philosophy of life. I wanted to grow my own food and have the freedom to work for myself, and for nature."*

In Indonesia, Adzhani explains, farmers don't want their children to follow in their footsteps, because that usually means remaining poor. "But farming doesn't need to be growing rice in the open land—farming can be many things," she says. "From your farm, you can provide education for people, or cook creative food from the produce you grow. I just wondered, 'What if I can do something different by becoming a farmer?'"

Today, she is the founder of her hometown's first rooftop farm, mingling three different farming methods that all come with their own challenges—no simple task, especially when the local ecosystem and climate are a world apart from the

European environment where she studied. In Konekroot's greenhouse, she works with aquaponics and hydroponics, while in the open air, both on the rooftop and downstairs in the backyard, she practices soil-based agriculture. Through these three systems, she is able to produce various vegetables and herbs, ranging from pak choi to kale. "I want to sell produce at affordable prices. Kale is very expensive here; you can only find it at the supermarket. We grow Chinese vegetables like kailaan, carrots, many kinds of spinach different to what Western people eat, local veg that is hard to find in the traditional market, and lettuce, even if it doesn't grow well in the heat."

*"Farming can be many things. I just wondered, what if I can do something different by becoming a farmer?"*

Starting her farm from scratch, and working with some methods for the first time, Adzhani's is a trial-and-error process. "I had to make the soil lighter because I had quite a few people coming onto the rooftop and it was too heavy, so I now use rice hulls," she explains. "With the fish in aquaponics, I failed a lot, and I cried a lot." No chemical pesticides are used on the farm; instead, natural remedies are applied, including a natural insecticide made from garlic. "Many pests come between the dry and the rainy season," Adzhani says. "We are close to the equator, and so the temperature doesn't change much. What changes are the seasons. Because of climate change, we don't have exact times for the dry and rainy season like we did before, when it was exactly six months. This makes it more difficult for farmers to work."

Adzhani's sustainability principles touch every facet of her business, from the soil blend she uses—made from rice hulls, an agricultural by-product—to the ways in which she recycles water or manages her surplus. If the flowers she cultivates aren't sold fresh, she dries them and sells them as tea. "I also make cold-pressed juices, because not many people want to cook vegetables, but they buy the juice. And sometimes I have too many seedlings and I don't have space to plant them, so I have a catalogue of seedlings available to purchase too."

In her quest to provide more opportunities for women, especially for those without

*1] Konekroot founder Rahma Nur Adzhani (far right) teaches children about farming as an outdoor activity (p. 147)*
*2] Harvesting micro cherry tomatoes from Konekroot's rooftop garden (top)*
*3] Meli inspects the honeydew plants for pests inside the greenhouse (p. 149 top)*
*4] Adzhani hosts the first open house and workshop, during which participants prepare a meal together with produce grown on the farm (p. 149 bottom)*

FARM
SHARE
LEARN
COMMUNITY
Farming

*5] Adzhani maintains plants growing in the rooftop garden (p. 150 top)*
*6] The aquaponics-grown honeydew melons are often pollinated by hand (p. 150 bottom)*
*7] A visitor to the farm harvests cherry tomatoes (top)*

access to higher education, Adzhani and her team—herself plus two women employees—organize educational workshops and host visits on the farm. "If I ever have an opportunity to replicate this at a larger scale, I really want to employ more women who didn't have the chance to complete higher education," she says. "The activities we run here are focused on women; they want to learn about healthy food and how to farm, and when we do workshops, the majority of attendees are women. They have become the center of food and farming at Konekroot."

These events can be demanding to organize, but they bring Adzhani real pleasure, and the opportunity to socialize and make new friends. Providing healthy, nutritious food to her community has become her top priority. "Some people can't afford supermarket food, but they can come here," she says. "I think that's a small patch of food security I have created in this small community. They really want to enjoy food, but they want a more affordable price, and I can help with that."

Adzhani's next innovation will be a café-restaurant in the backyard: "I want to sell probiotic drinks, fresh vegetable juices, and salads, all made with the vegetables that I grow. I want to offer something that is different," she says. "We normally don't eat salads or raw vegetables here; we eat cooked vegetables. I want to offer vegetables that customers can pick themselves and then eat: that's new in my city."

*8] Adzhani harvests edible flowers grown in Konekroot's backyard garden (p. 152)*

*9] Co-worker Meli greets guests at Konekroot's open house, where they cook and serve food grown at the farm (top)*

# How to Dry Flowers at Home

*Drying flowers is a simple activity that can bring beauty and joy by incorporating natural elements in your home, while also allowing you to keep memories alive through flowers*

The three most important things to take into account when drying flowers are air flow, light, and time.

The essential tools needed are secateurs, strings, and a rubber band.

Deciding what type of flower to dry is important since some flowers tend to dry better than others. Getting familiar with nature is handy because the more suitable flowers for drying are those whose petals hold their shape and color for a longer period.

For drying hand-cut flowers, the time to cut them is crucial. According to Lily Bruder-Zal of Vanishing Point Farm, the best time to pick the flowers is at the peak of their life cycle, so that they are perfect to preserve, or toward the end of their life cycle, when they are already starting to

*An urban flower pop-up in Brooklyn by artist turned flower farmer Lily Bruder-Zal of Vanishing Point Farm.*

*Hanging bouquets of dried hydrangeas inside Bruder-Zal's pop-up.*

*Assorted dried flowers brought together in harmonious displays of floral artistry.*

*Bruder-Zal gathers bundles of dried fern seed pods.*

*Outside view of Bruder-Zal's flower pop-up in Brooklyn.*

*Creating a welcoming space with branches, berries, grasses, and other natural elements foraged from Vanishing Point Farm's surrounding landscape.*

*Bruder-Zal hangs bouquets of celosia flamingo feather from the ceiling of her pop-up.*

*Bruder-Zal collects and arranges harvested flowers at Vanishing Point Farm in Highland, New York.*

*Artful sprays of assorted dried flowers on display at Bruder-Zal's Brooklyn pop-up.*

*Creating decorative autumnal wreaths with branches and dried flowers.*

*Bruder-Zal weaves celosia flamingo feather into one of her elegant floral wreaths.*

decay a little bit, since that helps in the curing process.

The color tends to get lost when the flowers are exposed to light, so the best spot to put them to dry is a dark place. Total darkness is ideal, as long as there is good ventilation and no moisture.

To dry the flowers, combine them in small bunches and tie the base of the stems with a rubber band or string. Hang them upside down in a cool, dark place with good airflow, ideally close to the ceiling. A damp-free cabinet with good airflow can work too.

Bruder-Zal uses a rubber band to tie the stems and opens up a paperclip so it forms the shape of an "S." Then she hangs one end of the paperclip from the strings on the ceiling, and she uses the other end to hold the rubber band.

It's convenient to turn the bundles a couple of times during the process to prevent the same side from always being exposed, as this helps them dry evenly.

The drying process can last between two weeks and a month, the latter being the better option to stay on the safe side.

Once the process is completed, store the flowers in a dark, black container, to keep their color intact for longer, and in a cool, dry environment.

Hydrangeas, lavender, celosias, and any plant that has papery petals can be dried beautifully.

# LUCIO USOBIAGA

*An entrepreneur fighting to save farming traditions in the* chinampas *of Mexico City*

Lucio Usobiaga has the approachable voice of someone who is aware of being immersed in a mission larger than himself. In his quest to prevent the ancient farming traditions of Mexico City's *chinampas* from dissolving into oblivion, he connects local farmers with customers—including some of the world's most celebrated chefs—to keep this precious pre-Hispanic heritage alive.

The *chinampas* are a collection of artificial floating islands built with vegetation and mud where crops are grown. Located mostly in Xochimilco, in southern Mexico City, they are home to two percent of the world's biodiversity. Oral wisdom has helped to preserve this farming system since the time of the Aztecs, and today the technique stands out for its ecological value. "The are a great example of how humans can alter their surroundings while supporting biodiversity and culture," Usobiaga explains. "They are a UNESCO World Heritage Site currently at risk of collapsing due to demographic pressure, neglect, and contamination."

While most of his well-off friends studied medicine or law, Usobiaga transferred from law to philosophy, against his family's advice. Heading for a career in academia, he decided to launch a store selling artisan products to complement his income while studying for his master's degree. "It was then that I felt the impact of what was going on in Xochimilco—the importance of that place and the area's ongoing decline," he says.

*"The* chinampas *are a great example of how humans can alter their surroundings while supporting biodiversity and culture."*

The more time Usobiaga spent reading about agriculture and getting to know local farming families, the more he wanted to get involved. "Agriculture integrates many aspects that instilled a sense of purpose in me. First, farming in itself: the technique, the seeds, the seasons, and the outstanding beauty in it. Then, the farmers' hard work. We need networks of support for them. In Xochimilco there are only a few left, and there isn't generational renewal," he says. "I realized there were many issues and also immense potential. When I completed my master's, I decided to devote my career to this."

In 2011, he co-founded Yolcan, a company that connects consumers with *chinamperos*

YOLCAN

YOLCAN
68
21
56
24
67
95
89
YOLCAN
94

(the farmers who cultivate the *chinampas),* offering decent working conditions for the latter. Winning farmers over wasn't easy—to do so, he bought one of the artificial islands and worked with local farmers familiar with traditional growing methods. "The strategy was to lead by example," he says. "We started to grow vegetables following ancestral techniques and permaculture principles. When they saw that the system worked and we were selling our organic produce at a better price, more *chinamperos* joined."

> *"When working with flowers, you are reminded all the time that you are going to die, which is for me something very positive. It's very vitalizing; it gives you the energy to live your life right now."*

In buying the high-quality *chinampas*-grown produce, individual customers (through a veg box scheme) and high-profile chefs like Jorge Vallejo and Gabriela Cámara support both sustainability and the *chinamperos'* traditional way of life. Usobiaga has also launched different gastronomic experiences, including school visits and private dinners, that bring visitors to see the *chinampas* and sample the islands' produce for themselves.

In 2014, Usobiaga founded IAX (Agroecological Initiative Xochimilco), a nonprofit that empowers sustainability-focused, small-scale farmers in different ways, from sharing knowledge to freeing the waters and soils of the chinampas from damaging residues. Arca Tierra is Usobiaga's latest venture, founded in late 2020 after he moved on from Yolcan. The name might change, but the mission remains the same: to offer the guardians of the *chinampas* decent working conditions and the opportunity to continue the legacy of their ancestors while providing superb fresh produce to the residents of Mexico City, improving food security in the metropolis. "We all live on the same planet. It's our only home and we need to look after it. One of the best ways to do that is by eating sustainable foods and supporting the farmers who grow them. In that way, we can create a network of mutual support that also benefits the environment," says Usobiaga. "I see that more young people are interested in this field. I am not the only one; we are a movement that knows many things need to be fixed to ensure a more hopeful future."

1] *Lucio Usobiaga,* chinampa *advocate and founder of Arca Tierra (p. 161)*
2] *Lettuce, sorrel, and amaranth grows on the* chinampas *of Mexico City (p. 162)*
3] *Flowering* Brugmansia arborea *with beds of lettuce in the background (p. 163 top)*
4] *The entrance to the Chinampa del Sol in Mexico City (p. 163 bottom left)*
5] Chapin, *an ancestral technique in which crops are grown in mud squares, is still used to grow produce on the* chinampas *(p. 163 bottom right)*
6] *Usobiaga explores the red amaranth grown on the* chinampas *(p. 164 top left)*
7] *Appetizer made from colorful carrots grown on the* chinampas *(p. 164 top right)*
8] *Chef Joaquín Cardoso plates salads with his team at an event organized by Usobiaga to raise awareness of the* chinampas *(p. 164 bottom left)*
9] *Rafael Estrada delivers* chinampa*-grown aubergines to restaurants and shops in Mexico City (p. 164 bottom right)*

# A Dreamy Organic Urban Flower Farm

*Blooming in northeastern Paris, Plein Air is the first urban floral farm in the city*

Poppies, dahlias, foxgloves, and amaranths are just some of the over 200 species that farmer florist Masami Charlotte Lavault, a former industrial designer, grows at Plein Air, the first organic flower farm in Paris. Founded in 2017 and covering 1,200 square meters (12,900 square feet), it is located in the city's 20th arrondissement, next to the Belleville reservoir and behind the Belleville cemetery.

"When I quit design and began to grow food, the big difference—and what changed everything for me—is that instead of making use of resources, you try to preserve them," explains Lavault. "You try to save them for later because you know you will need them year after year." After time spent farming in Morocco and Great Britain, Lavault moved to Japan, where she learned about effective microorganisms, an idea developed by the Japanese professor Teruo Higa.

Back in her hometown of Paris, following a decade living abroad, she thought that finding a plot of land to launch her farm would be easy. "It took me three years to find something," she says. "It was very difficult. And at first, I wasn't even in Paris, but in the suburbs. This space was actually my last shot; I was so fed up with not finding a space! I had six big refusals. People would say, 'This is not going to work—flowers, who cares?' And so I said to myself, 'Parisculteurs is my last try.'" The latter is a public program aimed at incorporating more of the natural world into the urban landscape by giving a new life as gardens or urban farming spaces to vacant plots and rooftops.

Lavault's proposal was accepted, and today, her flower farm is her life. Without any financial backup, she juggles three jobs and works 90 to 100-hour weeks to make it financially viable. Aside from the farm, she works twice a week at Normandy-based farm Une Ferme du Perche, where she tends the flowers. "I think it's a very fruitful connection, to work at an urban farm and a rural farm," she says.

An important part of Lavault's job at Plein Air is welcoming people, from the general public who come to buy flowers at the weekend to curious journalists. It makes urban farming very different from being in the countryside, where the focus is exclusively on farming. "Here, there is always someone trying to visit," she says, "which is very cute, and the interest is moving for me, but also time consuming."

Flowers are of course ephemeral, fading in a week or so. But Lavault considers that daily reminder of her own mortality a brilliant perk

of the job. "It's very vitalizing; it gives you the energy to live your life right now," she says. Every day is unique on the farm, since the nature of the work changes month to month. In early spring and late autumn, the main activity is to sow and plant; high season means harvesting and selling, as well as making the field look glorious for the weekend sales; and winter is reserved for planning and computer work.

> *"When working with flowers, you are reminded all the time that you are going to die, which is for me something very positive. It's very vitalizing; it gives you the energy to live your life right now."*

"Scheduling is a huge part of my job. Except in August, I plant all the time. To do that, I need to plan where to plant and when. I plan a season ahead—I usually plan the upcoming year in the summer," she explains. "I call autumn a 'second spring' because I have so many seeds for next year, and I have to plan and plan, but with the whole season behind us, I am physically tired and a bit burnt-out mentally. But I still need to act as if we are super fresh."

While the administrative tasks are shared with another employee, Lavault works solo on the farm, where she practices three different approaches that work well together. Drawn from the ideas of Austrian philosopher Rudolf Steiner, she follows biodynamic principles; secondly she uses effective microorganisms year-round through her irrigation system, a method Lavault learned in Japan that activates native microorganisms in soil and water, maximizing their natural power; and thirdly, she uses compost tea, a natural fertilizer.

With over 20,000 colorful plants, the farm today is a world apart from the tangled patch of land Lavault discovered when she first stepped in. "There was nothing here other than wild plants left on their own for years," she says. "Just cleaning it all without machinery took me a year and a half." She currently grows up to 250 species, an unusually high amount considering regular flower growers tend to have 10 to 20 different plants. But the wide range makes sense from a business point of view. "My field looks a bit crazy. There are plants everywhere—it looks more like a garden than a farm, and I like it that way," says Lavault. "It allows me to have 20 different

*1] Masami Charlotte Lavault, flower farmer and founder of Plein Air, the first urban flower farm in Paris (p. 167)*
*2] Lavault uses a seed sifter to sift delicate brome seeds (p. 168 top)*
*3] Tending to* Briza maxima, *known as* amourette *in French, meaning "infatuation" or "crush" (p. 168 bottom)*
*4] Young industrial hemp plants are used as foliage for bouquets (top)*

flowers to use in a bouquet in one spot. It's very useful because you can offer diversity and a very cool palette to your customers." On Saturdays, she sells her flowers directly at the farm (she previously also delivered using a bike, but three accidents persuaded her otherwise), and she also runs regular workshops to complement her income.

In terms of the challenges she faces, Lavault cites climate change as the biggest. "We need to adapt. There are plants for which the climate was OK 50 years ago, but it's too hot for them now," she says. "So, most of the plants I have here shouldn't be watered all the time; they are Mediterranean plants. We will have to change the whole field toward plants that can really handle a lot of heat and dryness." The responsibility on her shoulders can also be demanding. "I have roughly 20,000 plants here that need to be taken care of. They are not wild plants that appear here magically: I have grown them, I imposed my vision on that soil and that ecosystem, so I need to play my role all the time. I cannot just quit because it's too hot."

Lavault doesn't sugarcoat the reality of her chosen career. "People really think it's a romantic, easy life, and it's not. It's a lot of physical work, a lot of responsibility, and that's what makes it beautiful too. People dream a lot, which is great, but I feel it's important to tell the truth: farming is very hard."

Tough as her profession might be, she doesn't plan to change careers anytime soon. "Even if I work until I'm 70, I only have 40 seasons. One season is one shot because every season is very different, so I don't think I will ever get bored."

*5] Garden tools, raincoat, sunhat, and other essentials in the garden shed (p. 170)*

*6] Lavault creates beautifully curated bouquets with the flowers she grows on the farm, including strawflowers, snapdragons, statice, and foxgloves (top)*

# Sustainability and Permaculture

*Care for the environment and for life on the planet are the foundations of agriculture that looks to the future*

Unlike industrial agriculture, sustainable agriculture and permaculture use agricultural methods and philosophies that have been around for millennia. At their core is the concept of farming in the present, for the future; that is, growing food following reciprocity techniques that benefit both the people and the finite natural resources they depend on, keeping soils alive and conserving freshwater, so that future generations can continue to make productive use of the land. Regenerative agriculture, which prioritizes restoring soil health with practices like no-till farming or using cover crops, sequestering carbon in the process, goes a step further toward these goals.

Permaculture originally referred to farming systems intentionally designed to replicate the functioning of natural ecosystems. The term was coined in the 1970s by Australian ecological pioneers Bill Mollison and David Holmgren to describe an "integrated, evolving system of perennial and self-perpetuating plants and animal species useful to man." In the preview of *Permaculture: The Documentary*, Holmgren wonders, "Why is our agriculture all composed of only annual plants that grow and die in one year, whereas in nature there is diversity? By understanding how nature designs things, we can create permanent agriculture and permanent culture in everything we do."

The term permaculture is also used to define a global movement of individuals who organize their daily lives in harmony with nature, informed by the three ethics of permaculture: earth care, people care, and fair share. But the core basis of permaculture—working in reciprocity with nature, observing and understanding its patterns and cycles—is the legacy of indigenous peoples across the globe who might not have a given name to it, but have protected

down the evaporation of water after mulching and by nourishing our soil with natural fertilizers such as nettles," says Jankowski. Instead of buying new tools, the team recycle what's available—including the iconic yellow courier transport bags, which have been used over the years to transport several tons of compost to the rooftop farm. Orange mail boxes—where envelopes are usually stored—have become multipurpose tools at the farm, used to house seedlings or store vegetables such as tomatoes or carrots, while post trolleys are commandeered to move bales of straw.

*Indigenous peoples have passed down for generations the knowledge on which the foundations of permaculture are based.*

and passed down for generations the knowledge on which the foundations of permaculture are based.

In Paris, permaculture projects are flourishing. Communauté Facteur Graine is an open-air, 900-square-meter (9,700-square-foot) permaculture farm located on the rooftop of a post office in eastern Paris. A laureate from the 2016 edition of Parisculteurs, the farm is now run by postal workers. "It has allowed citizens to discover that we can cultivate on rooftops, manage organic waste, and give reclaimed materials a new life," says Sophie Jankowski, the farm's president and long-time permaculture advocate.

The farm is guided by the three ethics of permaculture, sharing the beauty of nature and the rewards of working collectively. It has also adopted permaculture principles in its farming methods: "We place compatible plants together so that they support each other, and we reduce our impact by transforming waste into resources, by slowing

Following organic principles, the team works to enhance biodiversity, a core permaculture principle. On the farm they grow any vegetables that can thrive in 20 centimeters (8 inches) of soil, from aubergines to tomatoes or berries, and the produce is then sold at a weekly market for the postal workers downstairs, and occasionally at external farmers' markets. Activities are as seasonal as their crops and include sowing seeds in the greenhouse, making compost or nettle manure, and tending beehives.

*Urban farms approach sustainability and permaculture in different ways. Using repurposed postal items like mail crates (p. 175 bottom) or yellow courier bags (p. 173), tending their own seedlings (p. 172), and working with woodlice to increase soil fertility (top) are some of the methods employed at Communauté Facteur Graine. At La REcyclerie, hens and Indian Runner ducks help reduce restaurant waste, while guests learn eco-responsible values in a fun and positive way (p. 175 top left and right).*

L'HEUREUSE MENAGERIE
LES POULES DE LA RECYCLERIE
DE RESPECTER
LA TRANQUILLITÉ
DES POULES ET CANARDS
NE PAS ENTRER DANS
LE POULAILLER SVP

Le Potager
Pétanque
FERME URBAINE
JARDINS
OUVERTS

LA POSTE

Farming sustainably is a complex idea, but at its core it is defined by a stewardship of natural resources—using resources efficiently to produce food so that the capacity to continue producing food in the future is not compromised. Some common sustainable practices in agriculture include maintaining healthy soil, promoting biodiversity, and managing water adequately.

*"Communauté Facteur Graine has allowed citizens to discover that we can cultivate on rooftops and manage organic waste."—Sophie Jankowski, president of Communauté Facteur Graine*

La REcyclerie, founded in 2014 and installed along the old railway line at Porte de Clignancourt in Paris, is a space of experimentation devoted to eco-responsibility—their motto is: "reduce, reuse, recycle." La REcyclerie houses an urban farm, a green area where chickens and ducks roam freely, a food forest, a collective edible garden, a repair station, and a café-restaurant. Its over 700 members pay an annual €30 ($36) fee that gives them free access to the repair service and to the drop-off for kitchen scraps, which are processed into compost and returned to those interested. The space is run with a circular-economy approach and the farm is maintained by members with guidance from two employees.

According to World Bank data, in most regions of the world over 70 percent of the globe's freshwater is used for agriculture. Gardening with less water is paramount when growing vegetables with a sustainable approach. At La REcyclerie, they achieve this by using drop-by-drop irrigation, as well as buried clay pots. "It looks like an amphora. You fill it with water, and because it's made of clay, there is porosity," explains Olivier Fontenas, coordinator of La REcyclerie's urban farm. "The water can go through the clay and water the plants around it. A bit like a reservoir, it waters the plants around it just enough."

*La REcyclerie, installed along an old railway line, is a space of experimentation devoted to eco-responsibility.*

The space organizes dozens of different events, from DIY workshops to seed swaps and cooking classes, many with a focus on promoting environmentally friendly habits. "Sustainability is so important because that's how we will survive tomorrow," says Fontenas. "If we don't change our habits, we won't survive. I feel La REcyclerie is like a school of new life habits: we teach and show you how to do your waste, compost, gardening, circular economy—it's a school for learning how to be sustainable for tomorrow."

*The urban farm at La REcyclerie is a refuge of biodiversity, with an edible forest, vegetable garden, rooftop meadow with beehives, composting system, indoor plant jungle, and henhouse. By introducing nature into the dense district of Porte de Clignancourt, the initiative promotes the development of ecological corridors (p. 177).*

# How to Start Composting at Home

*Urban composting is an environmentally friendly task through which kitchen scraps become nutritious organic matter*

Four main components—organic matter, oxygen, moisture, and bacteria—are involved in composting, a process that decomposes organic matter and turns it into compost. With this method, organic materials like food scraps that were initially doomed to landfill are recycled into a material that can be added to soil to help plants grow.

Organic elements used in the composting process include a combination of brown materials like dead leaves, which provide carbon, and green materials such as fruit rinds, which supply nitrogen.

A common composting ratio is one part green and two parts brown material. Chopping and shredding larger pieces speeds up the process. A dry, shady spot near a water source is the ideal location for a compost pile.

*Colorful food scraps ready to be processed at BK ROT.*

*Breaking down food scraps in the compost bin.*

*Local residents can drop off food scraps and collect compost for their garden at BK ROT.*

*Community-donated plant and flower cuttings together with local food scraps.*

*BK ROT's operations assistant Shaq shovels compost into the trommel, which helps separate non-decomposed materials from the finer finished compost.*

*A Green Guerrillas sign explains the process at community and public composting sites.*

*A large pile of compost in its curing stage, before becoming ready to use.*

*BK ROT youth leaders Crystal, Azucena, and Brys turn the compost piles, a key step in the process.*

*Steaming compost: healthy organisms in the compost will be active and produce steam even in winter.*

*Sealing up large barrels of compost ("black gold") for local delivery.*

Community-supported BK ROT produces compost from locally sourced food scraps in New York City. Ceci Pineda, the executive director, explains that they turn the compost piles every two weeks for around nine weeks. With pitchforks and shovels, materials in the center are brought to the edges and vice versa. Then the compost spends four weeks curing. After that, it is sifted.

Oxygen is needed to support the bacteria in breaking down the plant material. Turning the pile helps supply oxygen and it allows the compost to be produced more quickly.

A considerable part of the composting work is done by bacteria and other microorganisms already present that, with the help of water and oxygen, break down the plant material into useful compost.

Heat is released by the bacteria as they decompose the materials and is usually concentrated in the middle of the pile. Waiting at least two weeks to turn it lets the center of the pile heat up, creating optimal conditions for bacterial activity.

A compost thermometer can help monitor the different stages. According to Pineda, they use a thermometer to make sure present seeds are sterilized and pathogens killed. To achieve that, an average metric is to keep the compost temperatures at around 55 °C (131°F) for three to four days.

# Food Justice and Community-led Development Go Hand in Hand

*Promoting healthy eating and farming across their Brooklyn sites, East New York Farms! empowers the community*

Located in eastern Brooklyn, a culturally diverse and underserved community, East New York Farms! (ENYF!) operates two farms and two community gardens that work to provide solutions to pressing food justice issues by promoting local, sustainable agriculture and community-led economic development. This pioneering project, founded in the late 1990s, sits under the umbrella of United Community Centers (UCC), a social justice-driven community center that has been serving the East New York neighborhood for over six decades.

"ENYF! is the only urban agriculture, food justice led project in East New York," explains project director Iyeshima Harris. "We have started multiple community gardens in the area and help community gardeners with the process of starting their own gardens." The initiative offers extensive support to new gardeners, providing them with seeds and the option to sell the produce they grow at the ENYF! farmers' market. It also offers training opportunities for gardeners in the form of weekly workshops that teach different gardening skills and techniques.

Advocating for healthy eating is an important part of ENYF!'s mission and the main focus of their community education program. "We hire community members and we train them—people who love food and want to learn how to cook," says Harris. "They then go out into the community and perform live demonstrations, and make recipes from the produce on the farm, or recipes from the farm's cookbook. They basically educate the community on healthy eating or different produce that could be used to help combat a disease they might have." It's a service that is particularly crucial in their neighborhood, which is a victim of the contemporary destructive pattern seen in many low-income urban areas: the harmful combination of lack of access to affordable, fresh produce and the abundance of nutritionally poor fast-food options, both major contributors to high rates of food-related health problems such as obesity or type 2 diabetes. According to data from the city's Department of Health and Mental Hygiene, East New York suffers one of the highest incidences of type 2 diabetes in the city, with almost 15 percent of the population affected.

The two farms operated by ENYF! are both about 2,000 square meters (22,000 square feet) in size. UCC Youth Farm was founded first, in 2000, and is powered mostly by teenagers enrolled in the internship program. Each year, 35 paid interns aged between 14 and 18 who are living or going to school in East New York work on the farms and community gardens

Beans
FARMS!

with assigned supervisors. Interns have the option to stay in the program until they go to college.

Over 70 different fruits and vegetables are grown at this urban biodiversity oasis, including many specialty crops that speak to the diverse cultural backgrounds in the community. "We grow things like karela, also known as bitter melon, callaloo and amaranth, long beans, okra and jute due to the culture of the neighborhood and also our staff," says Harris. "East New York is predominantly a people-of-color community, and we make sure that every crop that we grow reflects the community—we are very intentional about how we think about and work with the community."

*"We have started multiple community gardens in the area and help community gardeners with the process of starting their own."*

Activities the interns complete include workshops covering ideas around social justice, racial justice, and food equity. They learn farming skills and work at the farmers' market, as well as providing hands-on support at fellow community gardens. According to Harris, "the main aim of the program is to teach youth how to grow their own food, expose them to urban agriculture, and also create some form of economic stability for them and their family, having them do something positive in the neighborhood."

Interns also lend a hand at ENYF!'s second farm, built in 2015 and located in the Pink Houses public housing development. Managed by Kelly Guevara, all the produce grown there is distributed for free within the community. "We also distribute produce for free on a weekly basis in East New York to residents who cannot afford fresh, local, organically grown food," says Harris. "Our produce is priced reasonably—tomatoes are $2 a pound—so we make sure we are meeting the community's economic needs as well."

The neighborhood also benefits from ENYF!'s composting facility, which has a 24/7 drop-off site and processes food scraps collected from local residents, the farmers' market, community centers, and a food pantry linked to a church they work with. "We use the compost we make on the farm, or if folks need compost, they can reach out to us and we give it to them for free," Harris says.

*1] Gemma Garcia in her garden, the St. John Cantius Parish Community Garden, which is supported by East New York Farms! (p. 185)*
*2] The UCC Youth Farm in Brooklyn is mostly powered by teenagers enrolled in the internship program (p. 186)*
*3] Youth intern Ayman and garden coordinator Elise wash and bundle beets for the farmers' market (p. 188 top)*
*4] Elise and community garden organizer Mikayla weed in the greenhouse at UCC Youth Farm (p. 188 bottom left)*
*5] Harvested karela, also known as bitter melon (p. 188 bottom right)*
*6] Youth farm manager Jeremy Teperman tastes one of the first cucumbers of the season (top)*

K FARMERS' MARKE
E TABLE
by local gardeners.

Covid-19 has disrupted their routines, making farm accessibility a challenge. Some workshops have been moved online and their farmers' market, usually held on Wednesdays and Saturdays, now operates just once a week. "All the produce sold at our farmers' market is from our farms—youth harvest it themselves and sell at the market," says Harris. "Some of our community gardeners sell their produce there too. Most of our community gardeners are elders, so selling their produce at the market when they don't have a steady income, or a fixed income, is a subsidy for them to help with whatever bills they have to pay, or whatever they have going on."

In the long term, one of the main challenges for ENYF! is finding a way to grow more food. "The challenge is how to implement a long-term, sustainable food-access program. We are seeing the pandemic affecting us and our neighborhood, and we want to make sure that our community is getting the best access there is," Harris says. "I know we can't do every single thing ourselves, but as long as we are there for our community, that's what matters to me—as long as we can still provide that access."

*7] Youth intern Shevanne works at the weekly East New York Farmers' Market (p. 190 top left)*

*8] Squash (pictured), corn, and beans, known as the three sisters in Native American agriculture, are celebrated at UCC Youth Farm's Three Sisters Garden (p. 190 top right)*

*9] Harvesting collard greens at the UCC Youth Farm (p. 190 bottom)*

*10] The East New York Farmers' Market, which also invites local community gardens to sell their produce (top)*

# RON FINLEY

*The artist and "gangsta gardener" who instigated a change in the law to make South-Central Los Angeles greener*

A natural multitasker with a design background, Ron Finley is a self-defined "gangsta gardener" who wears many hats, from urban gardening advocate to artist and urban anthropologist. His mission: fighting against the unfairness of the current food system by changing people's mindsets.

Over 23.5 million U. S. citizens live in food deserts—defined as areas lacking access to good-quality, healthy, fresh food—an echo of the systemic racism infused in the country's food system. Finley often refers to his neighborhood, South-Central Los Angeles, where fast food is ever-present but healthy, fresh produce is rare, as a food prison. "You can grow things in a desert—it's alive. If you look at a prison, that's a whole other kind of situation," he says. "That's basically what we have here—some people call it food apartheid or food slavery. It's criminal. It should be against the law."

Since he couldn't buy healthy food nearby, Finley decided over a decade ago to grow it himself, sparking a global movement in the process. Initially, he converted an unused strip of soil between the street and the pavement next to his home into an edible garden full of life. Breaking city guidelines—the garden didn't meet planning regulations—entangled him in a legal battle with City Hall, warrant of arrest included. But the fight paid off; the law was amended and thanks to his efforts, other Angelinos can legally follow his example.

*"What we have here, some people call it food apartheid or food slavery. It's criminal. It should be against the law."*

Finley's 2013 TED talk on guerrilla gardening was watched by millions and led to public speaking opportunities around the world. His gardening MasterClass, released in 2020, has become a global sensation and the first MasterClass to be streamed in prison systems in California and New York. "No idea how many gardens have been inspired by me around the world—by the emails I get, probably thousands," he says. "From Greece to Hawaii to Compton to Virginia, I know it's happening. I call myself an urban anthropologist; I am showing people how they could change their lives. We are nature, we decompose like a leaf does, we need to be taken care of, just like plants do."

CYCLICAL

CYCLICAL
ASSEMBLAGE
B
M
E

PLANT
SOME SHIT

One of Finley's most celebrated quotes is: "Growing your own food is like printing your own money." He explains: "Food is a currency—it has an intrinsic value. Not only is the trade there, but the health that it gives us. It's best if you know where the food comes from, what is put in it, if it's not then sprayed or has chemicals in it, and there comes the freedom. If you are growing your own food, you are free from an oppressive society, from an unjust food system. If they don't offer healthy food in your neighborhood, you can have it because you can grow it yourself."

*"You cannot eat diamonds. Farmers are the true humanitarians, the people who grow our healthy food for us."*

From his lusciously green edible garden, The Ron Finley Project, which is open to visitors by appointment, he advocates for a world in which kids know their nutrition and communities grow their own food. Finley grows a variety of crops, from pomegranates to bananas, which he shares with friends and neighbors. "In the garden, there are lessons every day—it's Armageddon," he says. "Every day there is a new battle, from ants to rats eating all your crops. It's a war, but as things come, you deal with it. You have to be on defence but no day is the same."

Although shifting people's mindsets is a challenge, Finley is determined to change attitudes. "We don't think about how the food gets to the store, if the people who harvested it can afford it, or how many hours of back-breaking work in the field they go through to grow a beautiful lettuce," he says. "We don't see the value in soil that we should see. That's how we change that mindset—seeing the power in the freedom that is in soil."

His latest project is to build an urban garden in South-Central L. A. that will serve as an example of a nutritious fruit-and-veg Eden. We need an "ecolution," Finley says—an ecological revolution for the planet—and he hopes that the pandemic will shift our values system. "You cannot eat diamonds. Farmers are the true humanitarians, the people who grow our healthy food for us."

*1] "Gangsta gardener" Ron Finley surrounded by greenery in his Los Angeles garden (p. 193)*
*2] The garden has expanded into what was formerly a swimming pool (p. 194)*
*3] A wide assortment of gardening tools are kept tidy in containers (p. 196 top left)*
*4] Bananas need a long and sunny growing period of between nine to fifteen months to fruit, followed by a further two to four months to ripen (p. 196 top right)*
*5] Located in South-Central L. A., Finley's green oasis includes a diverse variety of plants, trees, and herbs (p. 196 bottom)*
*6] Plants blend with art under blue skies (p. 197)*
*7] Pomegranates are one of the many precious crops grown (p. 198 top)*
*8] Finely in his element, surrounded by greenery and art (p. 198 bottom)*

# A Brazilian City Putting Power into the Hands of Local Producers

*On the importance of growing food as a tool to transform urban communities*

BY CHEF MANU BUFFARA

Food is a tool for transformation; healthy eating is capable of changing a country. Urban agriculture projects can have an enormous impact in so many areas—economically, environmentally, culturally, and socially. They can promote sustainable development, provide food and nutritional security, cultivate the land in an ecological way, promote health, generate income for families, and help to bring communities together. They can do all this and rescue the ancient wisdom of food production in the process. Precious botanical knowledge passed down through generations teaches us so much about species of medicinal plants, their pharmacological properties and planting characteristics—all vital information that helps our societies to thrive.

In the Brazilian city of Curitiba—my hometown—3 percent of the land is occupied by rural activities, both planting and raising animals. There are 89 gardens with some kind of partnership with the city, 42 of them in municipal schools.

In some gardens, you'll see beds marked with a colored stake that indicates they have a much larger variety of produce than others. These are part of the Horta do Chef project, an initiative I developed with Curitiba's City Hall to encourage producers to sell part of their harvest to renowned restaurants in the city. We also develop educational activities for these communities, teaching planting techniques and vegetable diversity, and share recipes that encourage full use of the produce—so often, nutritionally rich foods are wasted because people don't have the knowledge to put every element to good use.

For city halls, these types of projects are highly advantageous: they generate jobs and a steady supply of food for day-care centers and local schools. In Curitiba, 30 percent of the food supplied to schools comes from urban agriculture, promoting health through organic, pesticide-free food. (The use of pesticides in urban areas is prohibited in Brazil—somewhat ironic considering this country has championed the use of pesticides worldwide for the last 10 years.)

Urban Farm of Curitiba, a farm recently opened in the city, has become an unprecedented space in Brazil dedicated to educating sustainable agricultural practices in cities. It aims to provide hands-on

experience of the main stages of the food cycle, from the simple planting of seedlings to the preparation of food for conscious and sustainable consumption.

Another successful initiative is the The Honey Gardens program, which works to rescue pollination through our native stingless bees. It installs hives in myriad public spaces including parks, squares, schools, and community gardens. At my restaurant, Manu, we have five hives on our facade and are proud to contribute to the ecological awareness of our customers and residents in our neighborhood.

We hold in our hands the power to transform communities through food. Urban agriculture is not just a one-off project; for me, it is life, expression, and movement.

*Chef Manu Buffara's work focuses on nutrition, the environment, and fighting food waste. Over the last few years, she has built dozens of urban gardens for communities to grow their own in her native Curitiba, Brazil.*

# Farm to Fork

*By connecting to gastronomy and offering the complete journey from soil to table, some urban farmers take their love of good produce beyond growing food*

Organizing farm dinners and growing crops for their own gastronomic ventures help these entrepreneurs cement their ideals of fresh produce, provenance, and sustainability while offering their communities an alternative to the industrial food system. Furthermore, when growing their own, chefs and urban farmers have the flexibility to decide which crops to focus on, and, climate permitting, this fosters a valuable opportunity to cultivate culturally diverse foods that might be hard to find elsewhere.

De Stadsgroenteboer is a 4,000-square-meter (43,000-square-foot) farm located on the western side of Amsterdam. Founded by five friends who met while studying at the University of Gastronomic Sciences in northern Italy, the team encompasses three nationalities—Dutch, Swiss, and Colombian. Everyone in the group has different skill sets; they are gardeners, bakers, cooks, scientists, cheesemakers, and food entrepreneurs—and all share a love of good food.

Around 90 percent of De Stadsgroenteboer's production is sold through a Community Supported Agriculture (CSA) program. A CSA is a partnership through which customers commit to buying the produce grown by farmers for the season—in the case of De Stadsgroenteboer, from May to December. Through a tier-pricing system (the team sets a minimum and maximum price range and customers choose how much they pay within that range), they are exploring a model that is affordable for everyone in the city while juggling their own costs, which are very high.

On the farm, sustainability is a priority. De Stadsgroenteboer grows over 60 different vegetables, from artichokes to pak choi and herbs, following environmentally friendly methods. "We want to farm in a sustainable way. We are reading about it and experimenting all the time," says Julia Crijnen. "We follow organic rules; we don't use pesticides, herbicides, or chemical fertilizers. We really try not to disturb the soil—no plowing, no moving of the soil—and we try to keep the ground covered for the longest period. A lot of what we do is focused on maintaining soil life."

*"No plowing, no moving of the soil—we keep the ground covered. A lot of what we do is focused on maintaining soil life."*
*—Julia Crijnen, co-founder of De Stadsgroenteboer*

The farm also sells its produce to Amsterdam restaurant De Witte Zwaan, and they plan to explore further collaborations in the future. During lockdown, they created an extra vegetable scheme in order to sell produce that would have been destined for the restaurant.

Coordinating dinners in a working farm like De Stadsgroenteboer without a dedicated space for guests or cooking is a challenge with multiple issues to factor in, from storing the food that will be served to installing a temporary kitchen to cook it. The unpredictable Dutch weather didn't help their first farm dinner: "We really like doing it, but the way the farm is set up now—it's hard work because we have to re-organize the farm," says Milo Buur. The initiation ritual for diners upon arrival includes a short farm tour followed by a discovery walk where each guest finds little snacks. Patrons pick basil and cherry tomatoes from the greenhouse that are then put together with mozzarella to make one of the dishes on the menu, composed mostly from ingredients the farm has a surplus of that week.

Farm dinners are highly demanding, but the team has found other ways to demonstrate their cooking skills and share their love of exceptional produce with guests. "We have a harvest party, a spring party, and other events where we make food, but it's more casual and not like a dinner," says Crijnen. "That works well too—it's a lot of fun to do and less pressure."

Some farms have found success in creating a permanent space to welcome patrons, as in the case of ØsterGRO, the first organic rooftop farm in Denmark. Founded in 2014 by Kristian Skaarup, Livia Urban Swart Haaland, and Sofie Brincker, their greenhouse rooftop restaurant Gro Spiseri originally offered only dinner, but the pandemic has persuaded them to expand their opening hours to breakfast and lunch too.

*"You can't grow enough food to sustain yourself in the city. We are very dependent on the rural landscape."*
*—Kristian Skaarup, co-founder of ØsterGRO*

The produce for the restaurant is sourced from local farms around the Copenhagen area, while the food grown on the 600-square-meter (6,500-square-foot) rooftop farm goes to their CSA members. They grow a wide range of crops and focus on quality,

*Farming and gastronomy are inextricably linked from soil to plate. Babajide Alao, owner and chef of TheCradleNYC (p. 202), has partnered with Edgemere Farm (see p. 40) to grow some of the produce he uses at his restaurant, where he cooks West African dishes such as salmon with Nigerian fried rice and baked broccoli (p. 203 left). Andres Jara, one of the five co-founders of urban farm De Stadsgroenteboer in Amsterdam, harvests corn for their CSA veg boxes (p. 203 right). Danish farm ØsterGRO is a pioneer in urban farm dinners with its rooftop restaurant Gro Spiseri (previously known as Stedsans) in Copenhagen (p. 205).*

Kohberg

not quantity. According to Skaarup: "You can't grow enough food to sustain yourself in the city. We are very dependent on the rural landscape, and that's where we believe our food should come from." It's a common reality for urban restaurants, for whom it's not possible to fully source their produce from urban farms or grow all the food they need themselves.

Agricultural education, including activities such as guided tours, volunteering opportunities, and explaining in detail about the provenance of the food on their menus, is often part of the offer of many farm-to-fork restaurant models.

Sometimes, farming their own is a chance for chefs not only to expand the range of produce available, but also to reconnect with their roots. That's the case for Babajide Alao, founder of TheCradleNYC, a food business offering West African dishes in Alao's local neighborhood in Queens, New York City. Alao discovered Edgemere Farm (see p. 40) in Rockaway when he first volunteered there as a teenager. He began a new collaboration with the farm in 2020, and he now grows around 35 percent of the produce he cooks with on the farm, sourcing the rest from local markets. "Growing up in Nigeria, there was a lot of fresh produce," he says. "I was really interested in how things grow. Our main source of food was things that we grew in our yard."

*"I miss home, I miss Nigeria, and the only way I connect a lot with back home is through the music and the food."*
*—Babajide Alao, founder of TheCradleNYC and ESO*

From spinach and scotch bonnet peppers to collard greens and tomatoes, Alao uses the produce he grows for cooking, and also to create the drinks he sells at ESO, his smoothie and juice bar in The Rockaways in Queens. With a focus on the cuisine of his native Nigeria, his menu includes *jollof* rice and dishes such as *efo riro*, a spinach stew seasoned with spices imported from Nigeria. "I miss home, I miss Nigeria, and the only way I connect a lot with back home is through the music and the food," says Alao. "That's why I understood that a lot of Nigerians miss having fresh, organic Nigerian food that tastes like home."

*ØsterGRO (p. 207 bottom and p. 208) is a green breathing space for Copenhagen, with hens that provide eggs for CSA members and transform food waste from the kitchen into natural fertilizer for the fields (top). Chef Alao prepares baked chicken marinated in Nigerian stew sauce, served with vegan* jollof *rice and kale (p. 207 top left) at his brightly colored restaurant TheCradleNYC (p. 207 top right).*

UBER
eats
The Cradle
@TheCradleNYC
BOARDWALK

# A Revolutionary Rooftop Micro-vineyard

*Based on a rooftop in Brooklyn, Rooftop Reds, the first micro-vineyard in New York City, is open for tours and tastings*

Brooklyn and terroir might sound like an odd pairing, but Rooftop Reds founder Devin Shomaker made it a reality. The winemaker created his business, an urban rooftop micro-vineyard with magnificent Downtown Manhattan views, "to take the viticulture vineyard management that I was learning as a student and apply it to urban agriculture practices that I was self-learning, just because of my interest in urban agriculture, accessibility, and bringing green spaces to cities."

Following successful entrepreneurial experiences in China and Washington, D. C., Shomaker made a drastic career change in 2012 by enrolling in the viticulture and wine technology program at Finger Lakes Community College in western New York State. "My whole senior project was a climatic and environmental analysis of what grapes I should grow in Brooklyn," he says. "So how to develop, grow, and propagate the best varietals for the Brooklyn urban environment that I was targeting was part of my studies."

In his last year as a student, he set up a small vine nursery on his brother Thomas's rooftop in Brooklyn, hoping it would survive the winter. Based on the scientific analysis of the region, the environment, microclimate, and what was going to work, Shomaker planted traditional Bordeaux varieties, which thrive in cool, maritime climates.

> *"Where the grapes are grown has a bigger impact on the wine than where the wine is made. I never liked the urban winery model, where production is made in an urban space."*

A successful crowdfunding Kickstarter campaign that raised just under $20,000, and a $450,000 investment from a Finger Lakes winery, Point of the Bluff, allowed Shomaker to move his 50 high-quality vinestocks from his brother's rooftop to his current site. Located in Brooklyn Navy Yard, a 121-hectare (299-acre) industrial park, Rooftop Reds opened to the public in April 2016.

At their 1,400-square-meter (14,800-square-foot) rooftop space, which encompasses the micro-vineyard, a tasting room, picnic tables, and hammocks overlooking the Manhattan skyline, Shomaker and his team combine their focus on wine with an

eclectic program of events, including movie nights, full-moon yoga sessions, and small private gatherings. They also partner with different restaurants to offer pop-up dinners and pizza delivery, as well as a light snack menu. (The fully open-air rooftop closes in the colder months.)

Today, 168 grapevines thrive in this rooftop micro-vineyard. A mélange of five of the classic red Bordeaux varieties—Merlot, Malbec, Cabernet Sauvignon, Petit Verdot, and Cabernet Franc—is grown in a custom-made planter box vineyard system. The soil in the 42 planters scattered across the rooftop was created as a collaborative effort; Shomaker partnered with Cornell University researchers to produce a specific soil for the propagation of grapevines. "We set the clay ratio and the pH so the nutrients will be most available to our grapevines, and we used recycled glass as the sand element and the humus percentage in the soil that is best practice for vineyard management," says the entrepreneur. "There was a lot of tweaking—it's very different from a typical vineyard. Our whole development is, 'What if we had the most controlled, most perfect soil.' Because we get to develop it ourselves."

1] *Founder and managing partner Devin Shomaker (left) and his brother Thomas, a partner at Rooftop Reds, in the Rooftop Reds vineyard (p. 211)*

2] *Classic red Bordeaux grape varieties are grown and harvested by hand (top)*

3] *After pruning and weeding the vines, bird netting is hung to protect the grapes (p. 213 top)*

4] *Harvest day at Rooftop Reds usually happens between late September and early October (p. 213 bottom)*

*"Where the grapes are grown has a bigger impact on the wine than where the wine is made. I never liked the urban winery model, where production is made in an urban space."*

Set 18 meters (60 feet) above the ground, the vineyard benefits not only from abundant sun exposure, but also from Brooklyn's maritime climate, which provides a healthy aeration that helps control fungal issues. The vineyard is tended following organic and biodynamic principles, and once harvested, grapes are transferred to their production facility in western New York State, where fermentation and barrel aging also happens. "Where the grapes are grown has a bigger impact on the wine than where the wine is made. I never liked the urban winery model, where production is made in an urban space. You're paying a lot of money for urban commercial space, which inherently makes the wine out of reach for the consumer, price-wise. I price my rooftop wine very high because it is the

PETIT VERDOT

*5] Viticulturist Stephen Scarnato examines the grapes for any diseases at the point of veraison (the onset of the ripening of the grapes) (p. 214 top left)*
*6] Petit Verdot grapes are traditionally used in top Bordeaux red blends (p. 214 top right)*
*7] On harvest day, ripened grape clusters are identified and any damaged clusters are discarded (p. 214 bottom)*
*8] Genevieve waters plants in a greenhouse growing herbs and veg used in vineyard dinners (top)*

ROOFTOP REDS
GRÜNER VELTLINER
NEW YORK | 2019
GEWÜRZTRAMI
ROOFTOP REDS
SPARKLING ROSÉ
NEW YORK | 2019
ROOFTOP REDS
DRY ROSÉ
NEW YORK | 2019

first of its kind, but the wine that I trade under my label Rooftop Reds in the Finger Lakes wine region is very economical and very reachable to the masses. However, the price of my regular wines, not the rooftop wine, would need to be dramatically higher if my production facility were in New York City."

The premiere Rooftop Reds vintage made exclusively from grapes grown in the rooftop micro-vineyard was released in 2017—just 110 bottles, each with a $1,000 price tag, reflecting the uniqueness of these Bordeaux Blend vintages. The 2019 vintage, a mere 22 cases, is branded as a collector's item and has a similar price. Nevertheless, patrons without deep pockets visiting the rooftop space can opt to taste the substantially more affordable regular Rooftop Reds wines. Made from grapes sourced from New York State's Finger Lakes region, there are over a dozen different wines to choose from, starting at $11 per glass. Their NYC delivery options include the regular range, priced between $20 and $32.

Apart from their own bottles, Shomaker and the team also offer a wide range of wines, beers, and ciders made exclusively across New York State, including wines on tap that are kind on the wallet. "Wineries have often been for the elite, for people with the means to travel long distances to wine regions, and that is not the point of why I opened Rooftop Reds," Shomaker says. "Let's give everybody a vineyard winery experience they can get the subway to."

Sustainability is one of the company's founding principles. New greenhouses have been added to the rooftop, where herbs and tomatoes are grown for the environmentally conscious sushi restaurant Rosella, in the East Village. "It's as farm to table as you can get," says Shomaker, who considers global warming the biggest challenge for his company. "Urban agriculture should be pushed to the forefront as a major investment point.

Real money should be invested in businesses like Rooftop Reds, which actually help the environment, improve quality of life, and increase the focus on local produce that reduces our carbon footprint on our consumer decisions. If those things don't happen, that's a major challenge."

*9] Rooftop Reds hosts events including pop-up dinners, wine and cheese tastings, pizza-wine-movie nights, and educational viticulture tours (p. 216)*

*10] Seating areas benefit from the shade of the vines and offer spectacular Manhattan and Brooklyn views (p. 218 top)*

*11] Rooftop Reds' Grüner Veltliner, Gewürztraminer, Sparkling Rosé, and Dry Rosé (p. 218 bottom left)*

*12] Vineyard pop-up dinner with East Village sustainable sushi restaurant Rosella (p. 218 bottom right)*

*13] Glasses of Rooftop Reds' vibrant Sparkling Rosé and Channing Daughters Ramato (top)*

# ROOFTOP REPUBLIC

*An urban farming social enterprise that aims to change the relationship city dwellers have with their food*

Making cities more livable through urban farming is at the core of Rooftop Republic's mission. Established in 2015, this social enterprise has transformed dozens of unused urban spaces into more than 60 productive farms across Hong Kong and China. With clients that include residential developments, hotels, office buildings, shopping centers, restaurants, and schools, their service-based business model is making the Hong Kong skyline greener while building community bonds along the way.

This innovative company was co-founded by three partners: CEO Andrew Tsui, COO Pol Fàbrega, and CMO Michelle Hong. Tsui grew up between Singapore and Hong Kong, and has a background in real estate and civil engineering. Fàbrega, originally from Barcelona, started his career working in the third sector, covering issues from human rights to education. Since he relocated to Hong Kong in 2012, he has become an urban farming pioneer with several awards under his belt. Hong is a born-and-bred Singaporean with a background in communications and advertising. When she moved to Hong Kong, where more than 90 percent of food is imported, she began to grow her own and became more aware of how the food system works. Now she devotes her time to promoting the urban farming movement.

"We envision a future of sustainable cities and communities powered by urban farming," says Hong.

*"The best way to educate someone on the value of food and its impact on the planet is to have them grow it themselves."*
*—Andrew Tsui, CEO of Rooftop Republic*

"We offer farming as a service, providing urban farming consultation, design, build, and management services for our clients. We also help transform underutilized areas into vibrant natural spaces, create sources of nutritious organic food, and engage and empower communities to lead a sustainable lifestyle. Moreover, we offer education through workshops, programs, and our urban farming academy."

Rooftop Republic's small team has a hectic daily schedule. Not only do they build urban farms (some of them tended by office workers who donate the harvests

POAD

to food banks), but they also "operate many of them, managing the farms and engaging and educating stakeholders on the process of growing," explains Fàbrega.

With thousands of high-rise buildings in Hong Kong standing taller than 100 meters (330 feet) and little space on the ground, it was only natural that they looked to the sky for solutions. "We looked around and found that many rooftops and terraces were underutilized, and that gave us an opportunity to bring the farms to the people, instead of the other way around," says Tsui.

*"We envision a future of sustainable cities and communities powered by urban farming."*
*—Michelle Hong, CMO of Rooftop Republic*

Considering almost 70 percent of the world's population will live in urban areas by 2050, changing the value that people place on food is crucial for Rooftop Republic. "Urbanization, globalization, and modern living have designed a way of life that diminishes important aspects of our lives and creates an on-demand culture where many things—including food—have become so commoditized that city dwellers have lost that connection to their source," says Tsui.

*"We operate many of the farms, managing them and engaging and educating stakeholders on the process of growing."*
*—Pol Fàbrega, COO of Rooftop Republic*

From their personal experience of becoming interested in growing food as adults, and their professional experience building urban farms and engaging with people from CEOs to children, Tsui believes that "the best way to educate someone on the value of food and its impact on the planet is to have them grow it themselves." Guided by organic farming principles, their programs address the conservation of both the planet and our health. "Growing food in cities brings the story of food back into the urban narrative, allowing people of all ages living in high-rise, densely-populated neighborhoods to have a personal connection with how food is grown and methods of organic farming," he says. "Community gardens and farms welcome people to grow food together, building community bonds and strengthening human relationships over food. We believe that every city dweller has a part to play in the future of food, and it starts with planting a seed, growing food with your community, and rethinking the value of food."

*1] Rooftop Republic co-founder Pol Fàbrega (p. 221)*
*2] Jones Lang LaSalle rooftop farm in Central District, Hong Kong (p. 222)*
*3] Rooftop Republic co-founders (from left): Andrew Tsui, Michelle Hong, and Pol Fàbrega (p. 223 top)*
*4] Rooftop Republic's subscription program offers access to planters and urban farming workshops (p. 223 bottom)*
*5] Volunteers join in harvesting seasonal organic vegetables for donation to a local food bank (p. 224 top)*
*6] Co-founder Andrew Tsui leads an urban farming workshop for children, who harvest carrots by themselves for the first time (p. 224 bottom)*

# A Pioneering Rooftop Farm with a Successful Business Model

*With three locations atop industrial buildings in New York City, Brooklyn Grange has created a revolutionary rooftop farming model*

With three locations covering 2.3 hectares (5.7 acres) across New York City, Brooklyn Grange is a cutting-edge commercial rooftop farm, considered one of the largest in the world. Founded in 2010, this innovative business has paved the way for others who aspire to establish successful for-profit farming ventures in prime urban locations.

"Our goals for Brooklyn Grange are to grow food, increase urban green space, and build community through urban farming education and programming," says Anastasia Cole Plakias, co-founder and chief operating officer at Brooklyn Grange. "We want to make a strong case that social enterprise and good business can be powerful drivers of positive change."

Together with Ben Flanner and Gwen Schantz, Plakias founded the first Brooklyn Grange farm in Queens over a decade ago. Today, she is well versed in the complex process that setting up each farm entails: securing a number of permits, assessing each building's structure to ensure it can hold the extra weight of a green roof, and having the materials, soil included, lifted via a crane or blower truck. "It was the support of people around the world and here in New York City that kept us going," explains Plakias. "What we realized was that people wanted to believe that this model was possible, to create flourishing urban green spaces that produce food in the hearts of dense urban centers."

All three Brooklyn Grange rooftop farms are situated atop colossal industrial buildings: their first location was the historic Standard Motors Building in Queens, followed two years later by a rooftop farm at the Brooklyn Navy Yard, and most recently, a site in Brooklyn's Sunset Park neighborhood, inaugurated in 2019.

*"Our goals are to grow food, increase urban green space, and build community through urban farming education."*

The soils used across the three rooftop farms are made from different blends. These include mushroom compost combined with additional organic inputs and lightweight, porous stones to ensure correct drainage and provide the trace minerals that vegetables need. Varying between 20 to 30 centimeters (8 to 12 inches) deep depending on the farm, the soil is part of a green roof system that

includes a drainage layer that can hold up to an inch of water in reserve. This capacity to retain water allows the green roofs to capture stormwater, which benefits urban sewer systems by preventing overflow.

Growing over 4.5 tons of produce per year in a city with such an abundance of food cultures, Brooklyn Grange's opportunities to cultivate unique crops are endless. "We have always been very fortunate to learn a lot from our community, our CSA members (Community Supported Agriculture, a weekly veg box membership scheme), the chefs with whom we work, and our farmers' market shoppers, who provide feedback about what crops they value most and have even on occasion brought us rare seeds and starts over the years," says Plakias.

*"It's really meaningful to be able to connect with our neighbors and food traditions through these crops that we grow."*

*1] Farmers Alia and Matt prep soil beds for planting in a corner of Brooklyn Grange's Brooklyn Navy Yard Farm (p. 227)*
*2] Brooklyn Grange's Navy Yard Farm offers panoramic views of the Lower Manhattan skyline (p. 228)*
*3] Harvested Sungold cherry tomatoes (top)*
*4] Brooklyn Grange farmers Michelle and Tivon cart harvest crates full of freshly cut and washed salad greens to be refrigerated (p. 231 top)*
*5] Farmer Angela harvests rainbow chard at Brooklyn Grange's Sunset Park Farm (p. 231 bottom)*

In response to the Covid-19 pandemic, Brooklyn Grange altered their crop plan, increasing their food donations to 20 percent of their total production. "Previously, we focused on high-density, quick-yielding crops that have a fast turnover and don't take up a lot of space—things like salad greens and radishes, the majority of which we sold to chefs," says Plakias. "With the closure of many restaurants due to Covid-19, and the unprecedented need for emergency food relief, we shifted quite a bit of our crop plan over to crops that make more sense for families in the community who might be struggling to put food on the table."

Fundraising and charitable partnerships have enabled the farms to cultivate and donate these slower-growing, space-intensive, calorie-dense crops, which are often less profitable but more meaningful for their community. In Sunset Park, they grow vegetables that resonate with the significant Chinese-American population, such as mei qing choi, as well as pepper cultivars and herbs relevant to the cuisines of the Central American community. "It's really meaningful to be able to connect with our neighbors and food traditions through these crops that we grow," says Plakias.

Innovation and diversification are at the core of Brooklyn Grange's business. They have three selling channels: farmers' markets,

@massimomongiardo | 20

*6] The early morning sun slants across lettuce growing at Brooklyn Grange's Long Island City Farm (p. 232)*
*7] Farmer Donna harvests mustard greens at Brooklyn Grange's Brooklyn Navy Yard Farm (top)*

AVAILABLE FOR LEASE
1-888-888-6460
MASTER GLASS
HERBERT WOLF CORP

the CSA membership scheme, and wholesale, which includes restaurants and grocery stores. Additionally, the company organizes a myriad of events and workshops, from yoga on the roof to farm dinners and tour visits. In 2019, over 2,500 people joined the guided tours and more than 7,000 guests attended 150 events. They also offer consultancy, design, installation, and landscape maintenance services for private clients. One of their most celebrated projects to date is Vice Media HQ's abundant rooftop in Williamsburg.

*"We are a work in progress and each new season gives us another opportunity to learn from and improve on seasons past."*

"The Brooklyn Grange model was never to try and produce enough food to sustain entire cities, which isn't possible or practical," says Plakias. "Rather, we believe that in addition to crop production, farming on green roofs can provide tremendous value to an urban ecosystem. Offering environmental and social benefits is vital to the success of our model, and we do that in a variety of ways on the farm." Seeing the huge interest from urbanites to connect with their green spaces and agricultural knowledge, the farm's events have evolved over the years into robust educational programs, including a youth-specific one led by City Growers (see p. 71), a charitable nonprofit organization that engages thousands of kids in farm-based learning each year.

Plakias is aware that urban agriculture stands on the shoulders of giants such as Hattie Carthan and Liz Christy, co-founders of the Green Guerrillas movement. "Our work would not be possible without the work of those who came before us, often women of color, who recognized that their communities didn't have access to the educational benefits of green space, and often that healthy foods in America are a privilege and not a right shared by all." With their triple-bottom-line approach to business—focusing on planet, people, and profit—Brooklyn Grange is striving toward its vision of serving the community by creating more jobs, more green spaces, and more access to them for everyone. "We are very much a work in progress," Plakias says, "and each new season gives us another opportunity to learn from and improve on seasons past."

*8] Farmer Mark harvests green garlic at Brooklyn Grange's Sunset Park Farm (p. 234)*
*9] Farmers Seth and Julia harvest greens in the early morning at Brooklyn Grange's Long Island City Farm (p. 235)*
*10] Dinners give chefs and vegetables time to shine at Brooklyn Grange's Navy Yard Farm (p. 236)*
*11] Sky High Tomato Dinner at Brooklyn Grange's Navy Yard Farm, held in partnership with restaurant Joseph Leonard (top)*

*Seasonal worker picks tomatoes at NU-Paris (Nature Urbaine), Porte de Versailles, Paris.*

*Amaranth, rice, and lemongrass are some of the crops grown at the outdoor aquaponic Oko Farms (see p. 62) in Brooklyn.*

# Glossary

## Aeroponics

Aeroponics is the process of growing food in a soilless environment using a controlled fine mist of nutrients to feed the plants. No growing medium is needed for this closed-loop food production system. The naked roots hang suspended in the air inside a dark container, held in place by a horizontal board. As the plants grow, the roots are exposed to the nutrient-laden mist spray on one end, and the crowns to light on the other. Thanks to the lack of a growing medium, roots benefit from the extra oxygen, which tends to result in faster growth. It is vital to keep the roots hydrated, which is achieved by pumping water to the misting devices that allow for the mixed nutrient and water solution to be sprayed periodically, in the form of droplets.

There are two main types of aeroponic system: high-pressure aeroponics and low-pressure aeroponics. The main difference between them is the droplet size of the misty solution sprayed on the plants. As its name suggests, high-pressure aeroponics uses high-pressure, low-flow pumps and is consistently more expensive and more complex to build, while advantages include faster plant growth and the most reduced amount of water and nutrients use of any system. On the other hand, low-pressure aeroponics works with low-pressure, high-flow pumps and is easier to build and significantly more affordable.

Aeroponics thrives in an urban farming context, including indoor vertical farming, as it can produce large amounts of vegetables in tiny spaces, plants can be grown vertically as well as horizontally, and LED lights can be used to replace natural light. Leafy greens, herbs, strawberries, and tomatoes are some of the most popular crops to grow in aeroponic systems. The challenges for commercial aeroponic urban operations to become economically viable include large electricity bills and the high labor costs associated with metropolitan areas.

According to NASA—which has used aeroponic systems to grow food in space—aeroponics can reduce water usage by 98 percent, fertilizer usage by 60 percent, and pesticide usage by 100 percent.

Aeroponics is widely used in indoor vertical farming, and also by aficionados who build their own DIY versions of aeroponic systems at home. The elements to build a basic aeroponic system include an individual tank for the nutrient solution, an irrigation system, a water pump, and an enclosure that can maintain humidity levels while preventing light from reaching the roots.

Aeroponic farms can take many forms. AeroFarms, which was included in *Time* magazine's best inventions of 2019, use their own patented aeroponic technology in fully controlled indoor vertical farms. Agripolis (see p. 118), meanwhile, has multiple farms on rooftops across Paris, with produce growing in vertical cultivation columns and trellises that give plants nutrients via a closed-circuit watering system. Their most prominent project is Nature Urbaine at Porte de Versailles, the largest urban rooftop farm in Europe.

## Aquaponics

Aquaponics is a symbiotic ecosystem that combines the farming of freshwater fish with the cultivation of plants in hydroponics. In this closed-loop system, fish are fed nutrients and the waste they then generate is broken down by beneficial bacteria, providing the nutrients the plants need to grow hydroponically. The plants in turn purify the water, making it suitable for reuse. It's a system that mimics ecological cycles found in nature, such as the natural process that rivers or lakes depend on to grow vegetation.

The advantages of aquaponic systems are wide and varied. This process uses up to 90 percent less water than traditional farming methods, mostly due to its closed-loop nature and the ability of the plants to act as a natural filter, making it possible for the water to be reused. Additionally, the system doesn't require watering the plants or frequent water changes, and it eliminates farming chores such as weeding.

The versatility of the aquaponic system means it doesn't need vast amounts of space to thrive, making it suitable for the limited plots available in urban farming. It's a system that can be used both in the home and on a large scale. By producing food in an urban setting, aquaponics cuts down food miles, helps to reduce carbon emissions, and improves food security and the resilience of food-supply chains. Aquaponics is also environmentally friendly, as pest-control methods rarely involve the use of harmful agrochemicals that could damage the fish. Furthermore, food can be produced year-round and crops generally grow faster than in traditional soil farming.

Maintenance is challenging in aquaponics. For the well-being of the fish, water should be kept at a consistent temperature—this varies depending on the species—which might incur high energy costs. Water should be tested daily to control important elements such as pH levels, which should be kept within a neutral range to prevent the fish from dying and the plants from not being able to absorb the nutrients appropriately.

Typical vegetables grown in aquaponics include lettuce, spinach, kale, and herbs. The size of the fish tank determines the type of fish, but usually tilapia, trout, goldfish, barramundi, and bass are among the most aquaponic-friendly. Some aquaponic systems are more scalable than others; the most frequently used in larger, commercial-scale systems is the deep water culture system.

Nowadays, aquaponics is considered an environmentally and economically sustainable farming method. In 2019, the European Parliament listed aquaponics as one of the 10 technologies that could change our lives—mostly for its ability to produce local food without the input of any chemical fertilizers.

Aquaponic systems for the home are widely available, and thanks to companies such as Belgium-based BIGH Farms, Europe's largest aquaponic rooftop farm, commercial aquaponic farming in cities is becoming a reality.

## Beekeeping

In its simplest form, beekeeping is the activity of looking after colonies of bees, which usually live in man-made beehives, in order to extract the honey they produce. Domesticated beekeeping has been practiced since ancient times. When the first artificial beehives, known as "skeps," were developed around two millennia ago, it was necessary to destroy the hive to harvest the honey, often killing the entire colony in the process. Nowadays, the practice is considerably smoother for the bees, and they don't perish during harvest.

Contemporary wooden hives are based on the observations of American apiarist Lorenzo Langstroth, who is credited with revolutionizing beekeeping when he discovered that bees wouldn't make a comb in a space tighter than 1 centimeter (0.4 inches), to avoid the blockage of passages approximately their size. Using this concept, known as "bee space," in the 1850s he designed a type of hive with suspended, moveable frames spaced exactly one centimeter apart from each other and from the box walls. In this way, frames laden with honey could be easily removed without damaging the comb or irritating the bees.

Langstroth's design meant beekeepers could also easily inspect the hives for disease, monitor the bees' health, and harvest honey. Langstroth-inspired hives are still the most popular hives for professional beekeepers and hobbyists today and can be identified as any modular beehive with vertically hung frames.

Bees work in tune with the seasons, with their outdoor activity happening in the warmer months. While spring and summer are peak times for harvesting honey, it is crucial to monitor the bees throughout the year to ensure they are healthy and well fed, particularly during the colder months.

Each beehive functions as an independent organism where a single queen bee cohabits with up to 40,000 female worker bees—the queen's own offspring. (In winter, the average hive population descends to around 5,000.) The drones' main job is to mate with the queen. They are evicted from the hive as soon as mating season is over.

Habitat loss is one of the main existing threats for bees. To thrive, bees need food; they feed on pollen and nectar provided by flowers, so the best pollinator habitats are those with a mixture of plant species blooming from early spring to late fall, delivering enough nutrition throughout the seasons.

Urban beekeeping has boomed in popularity in the last decade, with the number of registered beehives more than trebling in cities such as London and New York. With bees threatened by habitat loss, pesticides, and climate change, some researchers believe cities could become a refuge for bees. Rooftops and back gardens are the most popular spots for city beekeepers to install their hives. The challenge for urban bees, which can fly up to five kilometers (three miles) a day in search of food, is to find enough forage available to feed on. Danielle Knott, an urban beekeeper, pollinator habitat advocate, and founder of Big Dipper Apiaries, promotes planting pollinator gardens with regional, native species as

*Beekeeper Danielle Knott handles anise hyssop, a perennial herb with scented foliage that is ideal for a pollinator garden.*

*Knott inspects her hive at the Naval Cemetery Landscape in Brooklyn.*

*Farm manager Jeremy Teperman and interns Armani and Hele decide what produce will go to market and what will be composted at UCC Youth Farm* *(see p. 184).* *The farm also has plots for community gardeners to grow in.*

a critical move toward providing food resources and habitats for bees.

## Community Gardening

Community gardeners have tended small plots and grown food and flowers in cities for centuries. From Tokyo to London, these gardens have become part of the cultural heritage of cities.

Community gardens are convivial spaces where different people from all walks of life grow food and plants together in a green urban environment. Some gardens are organized as community farms, with no individual plots at all, similar to urban farms, and their main activities are local food production and agricultural practices on land shared by a community of farmers. Oftentimes, urban allotments, which are large plots of land divided into smaller areas used to grow fruit and vegetables for self-consumption, are settled in public land and rented individually for a period of time, for a nominal fee. In their early days, urban allotments were associated with underprivileged families. The first were created around two centuries ago, as a chance for working urban populations to escape the bleak housing conditions they lived in while growing their own food and enjoying some fresh air. In Germany, these gardens were initially called *Armengärten* (gardens for the poor). Today, they are known as garden colonies and are often located on the edges of cities, each parcel separated by fences and with its own shed.

In Britain, allotments are ingrained in the collective memory of the Second World War. The iconic "Dig for Victory" campaign urged Britons to use any spare land to grow vegetables as part of the war effort, and by 1943 there were over 1,400,000 plots under cultivation, as well as an unspecified number of home gardens. The number of vegetable gardens decreased after the war, but they rose in popularity again during the recession of the 1970s. At present, allotments in cities such as London or Bristol are in high demand, and representative bodies such as the National Society of Allotment and Leisure Gardeners, founded in 1930, offer everything from legal advice to seed discounts to their members.

In the last decade, perhaps due to a growing environmental awareness, the waiting list for allotments in cities has increased hugely and the competition to get hold of one is fierce. Nowadays, they represent an opportunity for hobbyist farmers from any socioeconomic background to grow flowers and vegetables of their choice, from common produce that is easy to find at supermarkets or farmers' markets to culturally significant foods that aren't so ubiquitous.

Community gardens are a great tool to breathe new life into neglected urban land and they often offer volunteering opportunities, a chance for urbanites to get closer to nature and benefit from regular physical activity. Nearby water access, a shed to keep tools safe, and good public transport links or parking space options are some issues to consider.

Berlin's Prinzessinnengärten (Princess garden) at Moritzplatz was established in 2009 in an unused space that had been vacant for six decades. Founded by the nonprofit Nomadic Green, it was originally based on mobile urban agriculture, meaning crops are planted in transportable beds that can be moved anywhere. In the meantime, a second location where even more people can participate launched in 2020.

## Floriculture

Flower farming, or floriculture, is the cultivation and marketing of flowering and decorative plants both for gardens and flower arrangements. As a branch of ornamental horticulture, an important part of floriculture is plant breeding. Flowers are largely produced in temperate climates, and floriculture is often considered a greenhouse industry, although many flowers are actually grown outdoors. Farmed flowers can also be used in the cosmetic, perfume, and pharmaceutical industries.

Floriculture can be a type of intensive agriculture, in which case the income per unit area is greater than any other branch of agriculture. Urban floriculture can cut down the carbon footprint linked with the transportation of fresh-cut flowers while providing a seasonal product and a striking counterpoint to surrounding inner-city concrete structures. Sustainable floriculture can help pollinators thrive in cities, far from the intensive agriculture practices that rely heavily on harmful pesticides.

The production of annual bedding and garden plants, potted flowering plants, foliage plants, cut cultivated greens, herbaceous perennials, and cut flowers is usually considered part of floriculture. The cut-flower industry alone was

*True Leaves Floral co-founder Jessica Balnaves in her backyard flower farm in Brooklyn.*

*Starts of sunflowers, zinnias, shiso, Jewels of Opar, and cornflowers.*

*Balnaves harvests flowers at Brooklyn's El Garden, one of the urban gardens where she cultivates her flowers and herbs.*

*Creating bouquets for a weekly local flower share.*

*Community food forests planted by the Philadelphia Orchard Project have a wide array of fruiting plants including medicinal elderberry.*

*Volunteers at a work party press cider from apples grown at Beacon Food Forest in Seattle.*

*Maintained by volunteers, Beacon Food Forest combines agroforestry and permaculture principles to create a diverse edible landscape.*

worth over $29.2 billion in 2020, up from $1 billion in 1988. The biggest buyers are the European Union and the United States, while the Netherlands plays a critical role in importing and then re-exporting 40 percent of flowers from around the globe. Cut flowers can be grown in a greenhouse or field, and the most popular blooms are carnations, chrysanthemums, and roses—the latter extremely in-demand since the Victorian era.

Flowers are a perishable product—for every extra day spent traveling, flowers lose 15 percent of their value—so they are transported quickly between continents, mostly from Africa and South America to the Global North, using a cold chain. Similar to fresh produce, all transport modes used with flowers, from lorries to planes, have refrigerated facilities to keep them fresh.

The success of small-scale organic floriculture businesses is often associated with high-intensity production techniques. These pay special attention to soil health and use methods such as succession planting, meaning when one crop is done blooming, another crop is ready to be transplanted. This technique allows farmers to increase their overall production by having two harvests in one season. Subscription-based bouquets and weekly flower shares, a form of Community Supported Agriculture (CSA), is a business model that some small urban floriculture businesses such as Brooklyn-based True Leaves Floral are successfully adopting.

As cities continue to sprawl, more attention is being paid to the conservation potential of urban centers. Inner-city green spaces with abundant native flowers could help prevent habitat loss while providing a suitable environment for the conservation of pollinators.

## Food Forests

A forest is a complex ecological system in which trees co-exist and interact with other living organisms. A food forest is an edible landscape designed to incorporate the production of plant-based food, purposely integrating comestible elements in a broader, semi-natural ecosystem full of life, which not only provides yields for human consumption, but also habitats for wildlife.

Creating a food forest requires thoughtful planning and hard work at first, but once the different living elements are well established, less and less maintenance work is needed. Eventually, the aim is that food forests become mostly self-sustaining over time.

Borrowing from permaculture design concepts, food forests epitomize a sustainable and self-sufficient approach to food gardening. Horticulturist Robert Hart pioneered a seven-layer planting system that mimics natural ecosystems, based on his observations of the relationships and interactions of his forest garden. By intercropping vines and climbers with mature fruit trees and smaller trees, bushes such as berries, perennial vegetables, and edible plants that spread horizontally with tubers that grow underground, a whole new resilient agroecosystem made of mutually beneficial elements is created.

An important environmental benefit of urban food forests is their ability to increase our resilience to climate change; they can help regulate temperature and reduce the urban heat island effect, improve air quality, and provide a green space for everyone in the city to enjoy.

Local residents can interact with food forests and pick their fill among the edible fruits, nuts, and vegetables available. By making ready-to-be-collected food accessible at no cost, food forests have the potential to improve local food access in urban settings. Challenges to a successful food forest include pests, volunteers losing interest, and visitors over-harvesting.

Beacon Food Forest (BFF), located in Seattle, is a natural space with an emphasis on food access and community. The nonprofit Food Forest Collective acts as a steward of this edible landscape, which is handled by volunteers. The food grown at BFF is available for public harvesting and also donated to local food banks. "Our work includes dismantling the systemic racism that created food deserts and inequitable access to land ownership in south Seattle," explains community outreach coordinator Carla Penderock. "Access to healthy food shouldn't depend on your ability to pay or work for it, and shouldn't be limited by your race, ethnicity, or ZIP code."

According to co-founder Glenn Herlihy, the most challenging part of creating BFF was convincing the city of Seattle to take them seriously enough to grant BFF use of the land. Volunteers can partake in different ways, from helping out at weekly work parties to teaching a class. Anyone can visit and forage from the site, which has doubled in size since it was first planted in 2012, and still has room to grow.

## Homesteading

Self-sufficiency is the main goal of homesteading, a concept that prioritizes growing your own food in your own space and avoiding the reliance on others for nourishment as much as possible. At the heart of homesteading is the desire to lead a more sustainable lifestyle, whether that involves installing solar panels in your home, converting your backyard into a vegetable garden that provides most of your food, preserving seasonal produce, or growing some herbs on the window sill of your apartment.

Some might argue that homesteading is a way of life, a mindset that explores the ideas of resourcefulness and self-sufficiency, seeking to produce food wherever you are, and in the available space you have, as tiny as that might be. Homesteading tends to be physically demanding, as most tasks are labor intensive. Some urban homesteaders add animals to their repertoire; chickens and bees are typically some of the most popular in an urban setting.

An excellent example of successful urban sustainability is The Urban Homestead, a pioneering project founded in 1985 by Jules Dervaes in Pasadena, California, and continued by three of his adult children following his death in 2016. Over more than three decades, the Dervaes family have cultivated their 370-square-meter (4000-square-foot) garden, expanding their production to all available areas in their property, from paved zones to vertical surfaces. Today the homestead grows around 400 different vegetables, herbs, fruits, and berries, raising over 2,700 kilograms (6,000 pounds) of food annually. Using a growing method the family calls "square-inch gardening," many plants are packed closely together, mimicking how plants sprout in nature.

Over the years, the Dervaes family have become self-sufficiency overachievers, not only growing over 90 percent of what they eat, but also living mostly off the grid in terms of energy, by producing solar electricity and using homemade biodiesel to fuel their pickup truck. Eggs, milk, and manure are provided by their goats, chickens, and ducks. In addition, they sell a percentage of their organic produce to restaurants and customers.

In the United States, the homestead movement historically supported the free ownership of land in the American West for people willing to settle and cultivate it. Around 109 million hectares were distributed under the 1862 Homestead Act—usually in the form of 65-hectare plots. The only requirement for applicants was that they be at least 21 years old or the head of a family, which made single women and people of all races eligible, on paper. The last claim made under the act was granted in 1988, for a parcel of land in Alaska.

## Hydroponics

Hydroponics is a technique of growing plants without soil using a mixture of water and mineral nutrient solution to feed the plants. It's possible to cultivate the plants without a medium, but often growers use hydroponic-friendly media to support the roots of the plants and improve water absorption. Popular options include coconut coir (a shredded product made from coconut husks), sand, peat, and gravel.

Saving vast amounts of water is one of the benefits of hydroponic systems, which can use up to 90 percent less water than traditional field crop production. Plant yields are also consistently higher, as the chemical elements that boost plant growth can be controlled. Potential challenges include expensive equipment, vulnerability to power outages, or the need for some technical expertise to prevent waterborne diseases that spread quickly and mistakes that, without the buffer that soil provides, can become fatal.

Hydroponics is suitable for urban agriculture, especially vertical farming, as available space can be utilized to its full capacity, both horizontally and vertically. Commercial operations often use hydroponics in a controlled environment, in conjunction with artificial lighting that provides plants with the energy required for development. Typical crops include tomatoes, cucumbers, flowers, strawberries, and herbs.

Hydroponic systems can be divided into closed and open systems. While in open systems the nutrient solution is only used once, then discarded, in closed systems the nutrient solution and water

*Vegetables and herbs grown and harvested from a New Yorker's rooftop terrace garden in Brooklyn.*

*Lufa Farms' location in Montreal's Laval borough grows Lebanese and English cucumbers in a closed-loop NFT system. They are delivered through subscription boxes.*

supply is recycled and reused, with adjustments made as needed.

There are common DIY options among closed-loop systems, including the nutrient film technique (NFT), which consists of growing plants in a sloped platform where a small portion of the plants' root systems is constantly exposed to a shallow, flowing nutrient solution, while the upper part of the roots has continuous access to oxygen. Deep water culture (DWC) is another simple system you can build yourself, where the plant's root system is suspended in a nutrient-rich reservoir pumped with oxygen to prevent the plant from suffocating.

The ebb and flow system is very adaptable and also easy to build at home. The plants live in pots on a tray, which is regularly flooded with a nutrient solution from a reservoir via a water pump, then allowed to drain. Lastly, in the drip system, the nutrient solution is pumped up to the top of the plant so that it drips down through the root system with gravity. Additionally, aeroponics is also sometimes considered a type of hydroponics.

With four rooftop greenhouses scattered across Montreal, Lufa Farms is an outstanding example of an economically and ecologically sustainable model for hydroponic urban farming. They grow dozens of vegetables, greens, and herbs, following principles of sustainable agriculture, such as recirculating water, conserving energy, reducing waste, and avoiding pesticides. Their total production is able to feed 2 percent of Montrealers with fresh veg.

## Indoor Farming

Indoor farming is a rising industry, often portrayed as a more sustainable alternative to conventional agriculture. It consumes notably less fresh water, and produces food more efficiently and closer to the final consumer, cutting down the environmental impact of transportation in the process.

One of the main advantages of indoor farming is that food is grown in a controlled environment, so crops are protected from the consequences of shifting weather patterns and can also be grown year-round. The environmental conditions, such as humidity and temperature, of each farming unit can be decided in exact terms. Typically, most food cultivated in indoor vertical farms obtains energy from artificial LED lighting, and is grown following methods such as hydroponics and aeroponics, in which the food grows without soil and feeds on a nutrient-rich water solution.

According to World Bank data, it is estimated that 70 percent of the world's freshwater is used for agriculture. The most notable sustainability strength of indoor vertical farming systems is that they minimize water consumption, using 95 percent less water than conventional farms. Thanks to a combination of artificial intelligence and advanced communication technologies, indoor vertical farms produce higher crop yields than conventional farming. When indoor farms are located in urban environments, especially when they sit within restaurants or hotels where the produce goes straight from farm to table, packaging can be removed altogether.

Another benefit of indoor farming systems is that they use less space—plants are cultivated in layers on top of one another, significantly reducing the surface area needed and helping to tackle land scarcity. Moreover, these innovative farms can be built in the form of modular units, adapting their size to the location where the farm will be installed, from miniature to floor-to-ceiling vertical units. The model in which indoor farming units are specifically designed to be placed in partner locations like restaurants or grocery stores is known as "distributed farming networks."

A major challenge for the environmentally friendly credentials of indoor vertical farming is its higher energy usage. The artificial lighting and climate control systems needed in indoor controlled growing environments mean that this farming system has a considerably high energy usage.

Crops that are currently economically viable to be grown in indoor vertical farms include leafy greens and herbs. The ability to grow produce with a quick turnaround is also advantageous when responding to changing market demands, as has been the case for indoor vertical farms in recent months due to the Covid-19 pandemic.

## Rooftop Farming

Space in densely populated cities is scarce and expensive—one reason why a growing number of financially successful urban farming businesses are settled on rooftops. Taking advantage of a space that already exists and is often underused, rooftop

*Infarm (see p. 122) employees harvesting aromatic herbs at the METRO Nanterre warehouse, the largest indoor vertical garden in France.*

*Farm manager Nick Pesesky tends to yellow oyster mushrooms in Smallhold's fruiting chamber in Brooklyn.*

*Harvesting rainbow chard at Brooklyn Grange's Sunset Park Farm (see p. 226) in Brooklyn.*

*Snug Harbor Heritage Farm grows their tomatoes (including heirlooms like Brandywine) in a "high tunnel," a plastic-covered house without heaters.*

*Farm manager Ezra Pasackow harvests lettuce at Snug Harbor Heritage Farm in Staten Island, New York.*

farming contributes to making cities greener by cultivating fresh produce on top of buildings.

There are different approaches to rooftop farming depending on various factors, including the main goals of cultivation, whether the farms are commercial or community-run operations, and the growing methods used, which can range from soil-based farming to hydroponics or aquaponics. Climate is always a decisive factor in outdoor farming, even if greenhouses are used.

The structural capacity of the building is key to determining how much weight the rooftop can support and therefore what type of agriculture is viable. While industrial buildings might be sturdy enough to accommodate the extra weight of a greenhouse or 20 centimeters (8 inches) of soil with no issues, other buildings might present limitations.

Some rooftop farms produce vegetables in soil, just as ground-level farms do. Since both the space and soil depth are limited, rooftop farmers tend to choose fast-growing, shallow-root crops like lettuce or radish. Environmental factors such as sun exposure or winds whipping between buildings can become crucial just a few meters above the ground.

Greenhouses can be useful for various farming systems. These structures allow rooftop farms located in regions with harsh winters to grow vegetables year-round. Rooftop hydroponics, which involves growing food without soil in a mixed nutrient and water solution, often relies on greenhouses for protection from the elements.

Setting up a rooftop farm can be arduous, especially if there are no lifts or cranes available to take up the soil and equipment needed. Green roof systems, which include root barriers to prevent the plants' roots from penetrating the roof, are usually laid down before the soil, and drainage solutions hold excess water from heavy rainstorms. Daily tasks have to be performed without mechanization, and so labor costs are typically higher in rooftop farming.

Both green roofs and rooftop greenhouses are environmentally beneficial in a number of ways. They can produce locally grown fresh produce and provide noticeable energy savings for residents, as they keep dwellings cooler than bare roofs in the summer and trap heat in the winter. Green roofs can also play an important part in rainwater management, reducing the risk of flooding by preventing the existing drainage systems from overloading.

When open to the public, rooftop farms can become brilliant spaces for community building and bringing urbanites closer to nature. They can also provide unique educational opportunities for city dwellers—especially children—to learn more about how food is grown.

## Soil-based Farming

The term "agriculture" is most associated with a soil-based method of farming. Soil is the very foundation of agriculture, providing the nutrients and minerals crops need to develop.

It is believed that humans domesticated the first plants around 10,000 years ago, at different times and in different spaces. While growing methods have evolved hugely, a healthy soil is still paramount for successful ground-based soil agriculture. A neglected, agrochemical-dependent soil contributes to the degradation of its microbial ecosystems, and the resulting decrease of nutrients available will negatively affect plant growth.

Throughout history, most ground-based farming has happened in native soils, and that is still predominantly the case today. The type of soil is a key factor in determining what cultivations work best. Major ground soil textural classes include sandy, silty, clay, and loam soils. Ideal soil care includes applying organic fertilizers such as composts or manures—organic matter that will support the aeration and looseness of the soil. However, contemporary non-organic agriculture relies on synthetic fertilizers—notoriously nitrogen—to grow food.

One of the processes that prepares soil for cultivation is tilling, a breaking up and turning over of the soil. Tilling became popular because it allowed farmers to plant more seeds, faster. Mechanical agitation such as stirring or digging, or human-powered picking or shoveling, are common tilling practices. Soil tilling—especially deep tilling—contributes to soil degradation in the long term because it leaves the topsoil bare, making the remaining soil more exposed to erosion by wind and water while disrupting the soil's structure and compromising its capacity to absorb water and nutrients.

Soil erosion, defined as the accelerated removal of topsoil from the land surface through water, wind, and tillage, was considered one of the major soil threats identified in the UN's 2015 "Status of the World's Soil Resources" report. Regenerative agriculture farmers frequently adopt a no-till approach, which helps prevent soil erosion and benefits microorganisms vital to soil health.

Soil is a steward of fresh produce and a healthy environment. Snug Harbor is a superb example of sustainable agriculture with an emphasis on building healthy soils. Located in New York City's Staten Island, this one-hectare heritage farm produces fruit and vegetables for restaurants and community members through their CSA program. Following sustainable principles, their low-till approach to farming focuses on soil health by using cover cropping, crop rotation, and compost use. They also support pollinator habitats by planting flowers that attract insects and pollinators to help them maintain a healthy balance of ecological diversity in the field.

## Underground Farming

As its name suggests, underground farming is the practice of cultivating food below ground. The vast majority of subterranean farms rely on NASA-pioneered energy-efficient lighting technology to replace the natural sunlight that plants need for photosynthesis. Mushrooms are an exception since they can be grown in dark conditions. Hydroponics, the practice of growing food without soil using a mix of nutrients and water to feed the plants, is often the cultivation method of choice in underground agriculture.

Energy costs are a key factor in determining which crops can be grown in artificially lit systems underground. Plants absorb the sun's energy through photosynthesis, converting carbon dioxide and water into biomass. The heavier the vegetable, the more energy is needed to produce it. When using artificially lit systems, this adds to the expense and to the environmental impact. In recent years, low-energy LED systems have included specifically calibrated lights that meet the exacting energy needs of each plant. Nowadays, only low-biomass, quick-turnaround crops such as leafy greens, or herbs such as basil or chives, are economically viable to be grown underground without natural light.

One advantage of underground farming is that crops aren't at the mercy of the weather, allowing farmers to have more control over the growing environment and to grow food year-round. Pests are also less recurrent than in outdoor agriculture, and as with any type of urban farming, produce grows nearer the final consumer, reaching them faster and at its freshest, cutting down carbon dioxide emissions from transportation. Reducing the environmental impact of substituting the sun's energy for artificial lighting is one of the biggest challenges for artificially lit underground agriculture, as is achieving a good airflow, which prevents pathogens from building up. The UN predicts that by 2050, 68 percent of the world population will live in urban environments. With substantial unused underground infrastructure in sprawling cities, some consider farming underground an alternative that will become more economically viable as LED technologies improve their efficiency and their cost shrinks.

London-based farm Growing Underground is an excellent example of subterranean agriculture. Located in a unique setting 33 meters (108 feet) below the streets of south London, they grow microgreens vertically using hydroponic growing trays, swapping sunlight for LEDs. Being so far underground prevents pests from entering the site and keeps the temperatures consistent all year round. "The biggest challenge of growing underground is the logistics, particularly as we are located within Second World War air-raid shelters that have a small lift shaft and a winding staircase. This causes challenges when bringing large pieces of equipment into the farm," says co-founder Richard Ballard. The farm has been open for tours since 2019.

*ECLO grows mushrooms and microgreens in a 3,000-square-meter (32,300-square-foot) underground farm in Brussels. Their unique production method uses an organic brewery's spent grains to produce their mushroom substrates.*

*Specialty herbs, microgreens, and edible flowers grown by Farm.One (see p. 132) in Manhattan, including dianthus, marigold, and blue spice basil.*

# Index

CASE STUDIES



FEATURES

Dave Bagnall Collection / Alamy Stock Photo
*p. 09 (top)*
United Archives GmbH / Alamy Stock Photo
*p. 09 (bottom left)*
Viennaslide / Alamy Stock Photo
*p. 09 (bottom right)*
FPG / Getty Images
*p. 11 (top)*
Richard Levine / Alamy Stock Photo
*p. 11 (bottom left)*
Valery Rizzo
*p. 11 (bottom right)*

**Sustainability and Permaculture**
Valery Rizzo
*pp. 172–177*

ACTIVITIES AT HOME

**Beekeeping**
Valery Rizzo
*pp. 112–115*

**Composting**
Valery Rizzo
*pp. 178–183*

**Drying Flowers**
Valery Rizzo
*pp. 154–159*

**Natural Dyes**
Valery Rizzo
*pp. 90–95*

**Small Space Growing**
Valery Rizzo
*pp. 44–47*

PROFILES

**Greg Kimani**
Nairobi, Kenya
*mwengenye2019.netlify.app*
Sven Torfinn / Panos for Argidius
*pp. 139, 140, 141 (bottom), 142, 143, 144*
Greg Kimani
*p. 141 (top)*

**Kat Lavers**
Melbourne, Australia
*katlavers.com*
Amy Piesse Photography
*pp. 49–53*

**Lucio Usobiaga**
Mexico City, Mexico
*arcatierra.com*
Leila Ashtari/@ashtariphoto
*pp. 161–164*

**Meredith Hill**
New York City, U.S.
*ms371.org*
Valery Rizzo
*pp. 77–80*

**Ron Finley**
Los Angeles, U.S.
*ronfinley.com*
Thomas Vickers
*pp. 193–198*

**Rooftop Republic**
Hong Kong, HKSAR
*rooftoprepublic.com*
Nic Gaunt
*p. 221*
Xaume Olleros for WWD
*p. 222*
Xaume Olleros / Bloomberg
*p. 223 (top)*
Courtesy of Rooftop Republic
*pp. 223 (bottom), 224 (bottom)*
Sarah Thrower
*p. 224 (top)*

GLOSSARY

**Aeroponics**
Giovanni Del Brenna
*p. 238 (top)*

**Aquaponics**
Valery Rizzo
*p. 238 (bottom)*

**Beekeeping**
Valery Rizzo
*p. 241 (top)*

**Community Gardening**
Valery Rizzo
*p. 241 (bottom)*

**Floriculture**
Valery Rizzo
*p. 243*

**Food Forests**
Jonathan H. Lee / subtledream.com
*p. 244 (top right and bottom)*
Courtesy of Philadelphia Orchard Project
*p. 244 (top left)*

**Homesteading**
Valery Rizzo
*p. 247 (top)*

**Hydroponics**
Courtesy of Lufa Farms
*p. 247 (bottom)*

**Indoor Farming**
Giovanni Del Brenna
*p. 249 (top left)*
Valery Rizzo
*p. 249 (top right)*

**Rooftop Farming**
Valery Rizzo
*p. 249 (bottom)*

**Soil-based Farming**
Valery Rizzo
*p. 250*

**Underground Farming**
eclo.be (photo: Arne Baensch)
*p. 253 (top)*
Valery Rizzo
*p. 253 (bottom)*

**Urban Farmers**
*The Now (and How) of Growing Food in the City*

This book was conceived, edited, and designed by gestalten.

Edited by Robert Klanten and Andrea Servert
Co-edited by Valery Rizzo

Written by Mónica R. Goya
Essays by Annie Novak (pp. 20–21), Chris Bolden-Newsome (pp. 54–55), Nic Dowse (pp. 102–103), and Manu Buffara (pp. 200–201)

Editorial Management by Anna Diekmann

Photo Editor: Valentina Marinai

Head of Design: Niklas Juli
Design and Layout by Johanna Posiege
Illustrations by Gino Bud Hoiting

Typefaces: Baskerville by John Baskerville and Domaine by Kris Sowersby

Cover image by Valery Rizzo
Back cover image by Courtesy of ØsterGRO

Printed by Schleunungdruck GmbH, Marktheidenfeld
Binding by Hubert & Co. GmbH & Co. KG. BuchPartner, Göttingen
Made in Germany

Published by gestalten, Berlin 2021
ISBN 978-3-96704-006-7



*For more information, and to order books, please visit www.gestalten.com*

*Bibliographic information published by the Deutsche Nationalbibliothek. The Deutsche Nationalbibliothek lists this publication in the Deutsche Nationalbibliografie; detailed bibliographic data is available online at www.dnb.de*

*None of the content in this book was published in exchange for payment by commercial parties or designers; gestalten selected all included work based solely on its artistic merit.*

*This book was printed on paper certified according to the standards of the FSC®.*

*Valery Rizzo is an American portrait, food, and lifestyle photographer focused on urban life and agriculture. Her work has appeared in* The New York Times Style Magazine, Yes! Magazine, Télérama, D La Repubblica, Der Spiegel *among other publications.*